ENERGY
The Conservation Revolution

MODERN PERSPECTIVES IN ENERGY
Series Editors: David J. Rose, Richard K. Lester, and John Andelin

---

ENERGY: The Conservation Revolution
John H. Gibbons and William U. Chandler

STRUCTURAL MATERIALS IN NUCLEAR POWER SYSTEMS
J. T. A. Roberts

ENERGY
The Conservation Revolution

John H. Gibbons

AND

William U. Chandler

PLENUM PRESS • NEW YORK AND LONDON

Library of Congress Cataloging in Publication Data

Gibbons, John H   1929-
  Energy, the conservation revolution.

  (Modern perspectives in energy)
  Includes index.
  1. Power resources–United States. 2. Energy conservation–United States. I.
Chandler, William U., 1950-   joint author. II. Title. III. Series.
TJ163.25.U6G52   333.79'16'0973   80-28431
ISBN 0-306-40670-5

First Printing–March 1981
Second Printing–March 1982

© 1981 Plenum Press, New York
A Division of Plenum Publishing Corporation
233 Spring Street, New York, N.Y. 10013

Printed in the United States of America

Preface

We try in this book to provide a detailed but readable, technical but accessible monograph on energy in the United States. We treat energy as a multidisciplinary challenge and apply the standard tools of economists, physicists, engineers, policy analysts, and, some might claim, fortune tellers. We hope that it will be used in classrooms of various types, and read by the general reader as well.

That increased energy efficiency should be the first priority of energy policymakers is a conclusion, not an assumption, of our analysis. Many analysts have arrived at this conclusion while working separately on energy supply problems. The magnitude and scope of supply problems, primarily problems of high prices and environmental costs, lead one inexorably back to reducing demand growth as the first, most important step in *any* plausible energy future.

We examine, in some depth, why much of the past literature on energy still points, fallaciously in our opinion, to high energy consumption futures. This is in Part I (called "A Short History of the Future").

We devote one-third of the book (Part II) to energy resources, their internal and external costs, and the quantities of energy to be derived from these resources. This analysis provides a context within which the economic and social value of energy conservation options can be assessed.

The value of energy conservation is examined in Part III. We study the technical and institutional opportunities and barriers to success that this most important option presents and faces. Special attention is paid to the cost effectiveness of higher energy productivity and to methods of "delivering" energy conservation measures.

We owe many intellectual debts for this book: to the University of Tennessee for encouraging the book's preparation while the authors worked at the University's Energy, Environment, and Resources Center (EERC); to the many scientists whose work we have integrated and whom we tried to acknowledge in footnotes; to Roger Carlsmith, David Rose, Robert Bohm, Harold Federow, and Holly L. Gwin for careful technical review and advice; to Marvin Bailey for professional editing; to Joyce Finney of EERC and Louise Markel of the Institute for Energy Analysis for being creative (and patient) information specialists; to Diane Heflin, Joseph Simpson, James Butler, and Fay Kidd for contributing their talents to the preparation of the manuscript; to the Environmental Policy Institute, especially Louise Dunlap, for encouraging one of the authors (Chandler) to make finishing touches during his employment time; and to David Rose and Plenum Publishing Corporation for this series of books on energy, for which we have the privilege of presenting the first.

This book was largely completed while both of us were associated with the University of Tennessee. The opinions expressed here are not derived from, nor do they necessarily reflect, those of the University or the institutions with which we are now associated.

This is not a book on the whole energy problem — that is a task for someone braver or wiser than us. Our emphasis is on energy as it is *used* to help provide desired goods and services to people and organizations. We would shrink from saying, and vigorously deny, that the U.S. can resolve its energy problems purely by focusing on using energy more efficiently. Instead, we are persuaded that vigorous efforts on the energy supply side will be required even if we hope to maintain *present* production rates. But as energy prices inexorably rise, so does the incentive to use energy more productively. Fortunately, as we demonstrate here, there are many technological opportunities to do just that. Beyond the incentives to minimize cost, the imperative to conserve (to use wisely) also derives from consideration of other people — in other lands and in future years. It is to them as well as to the American energy consumer that we respectfully dedicate this book.

John H. Gibbons
William U. Chandler

Contents

Chapter Two: The Elements of Energy Demand

Chapter Three: The U.S. in World Society

Chapter Six: Conjunctures of Energy and Environment

Chapter Seven: Energy Supply Policy

PART THREE: THE CONSERVATION WELL
Introduction to Part Three, p. 157

Chapter Eight: Buildings—More Amenities, Less Energy

Chapter Nine: The Crisis and the Car

Chapter Ten: Industrial Sector Conservation

We are in the middle of the Industrial Revolution.
We had better be; we have much to set right.

—JACOB BRONOWSKI
The Ascent of Man (1974)

Prologue

The Conservation Revolution

P.1. In the Beginning

About 4.7 billion years ago a star blew up, because of gravitational collapse, and became a supernova. In ten seconds, with neutron densities of hundreds of tons per cubic centimeter, many of the heavy elements of the solar system, today's material heritage, were formed from lighter ones. Less than one hundred million years later, just a moment in astronomical time, the solar system condensed out of the debris of that explosion and formed the sun and planets. Roughly 4.3 billion years later (two hundred and fifty million years ago), the carboniferous period evolved on Earth with prolific swamps, lepidodendron trees, and giant reptiles. Sunlight was photosynthesized, then fossilized and stored under sea and rock. Meanwhile, radioactive elements released energy inside the earth, thereby creating earthquakes and volcanoes which slowly but elegantly segregated and concentrated mineral wealth.

Thus, all forms of energy with us today, whether derivable from gravitational, electromagnetic (chemical), or nuclear sources, were orginally derived from the weakest: gravitation. Tidal energy results from the conversion of gravitational energy in the earth–moon system. Gravitation holds together that steady-state thermonuclear fusion device, the Sun. The chemical energy in the organic bonds of coal, oil, and gas molecules is ancient stored sunlight. Wood, wind, and hydroelectric power are sunlight once removed. The heavy elements from which nuclear energy derives were created and the carbon atoms in this page and in the eyes that read it were scattered in that supernova.

Loren Eiseley's, "We are born of the light and of the dust of a star," and Walt Whitman's, "I believe every leaf of grass is the journeywork of the stars" are true, exactly.

P.2. Revolutions

While it may not be necessary to go all the way back to creation to begin an analysis of energy, more is called for than the usual, "Beginning with the Arab Oil Embargo of 1973. . . ." It is important to realize that there was very little comprehensive energy analysis before that historic event. It is also important to remember that the Embargo happened to coincide with an already planned major price increase. The Arab section of the Organization of Petroleum Exporting Countries (OPEC), playing a role similar to the Greek women in *Lysistrata,* cut oil production in protest of Western support of Israel in the Yom Kippur War at the same time as OPEC *in toto* hiked the price of oil by 66 percent. This event reversed the decline in real energy prices which had prevailed for decades. Oil cost more in the United States in 1950 than it did in 1970, and the same was true for electricity. The quadrupling of oil prices which followed the 1973 Embargo snapped energy prices back to their 1950 level. And because the price increase was so abrupt, it focused our attention dramatically on energy and forced us to ask basic questions about our ways of solving problems and our relationship to the rest of the world. Out of it has come a reassessment of traditional perspectives.

Revolutions, as much as anything else, change the way we perceive things. But a revolution need not be a rebellion. Rather, we would hope that the energy conservation revolution will become, in Thomas Jefferson's words, "the extraordinary event necessary to enable all the ordinary events to continue."[1] The invention of writing was such a revolution for it facilitated the keeping of records necessary for managing trade and agricultural surpluses in the emerging civilizations along the rivers of the Middle East. In Ancient Greece the revolutionary concept of court-applied justice replaced justice through revenge such as that chronicled in Aeschylus' *Oresteia.* But perhaps no revolution has affected the human condition as has the Industrial Revolution. Although the use of machines powered with nonhuman energy spans over many centuries, industry as a mode of life exploded in the aftermath of the Enlightenment of the eighteenth century. Making the Industrial Revolution possible, and perhaps making it necessary, were transformations of agriculture, banking, science, technology, trade, and commerce, all of which helped feed growing millions of people and freed, or uprooted, them to work in the new mills. It took a labor revolution, some would say a rebellion, to correct some of the worst abuses of industrialism, and the world today is divided ideologically over how to correct the remaining problems.

The problems are profound. The industrial world seems to be rushing lemming-like to a precipitous exhaustion of oil. Impoverished nations cling tenuously to deteriorating domestic resources and slowly lose their ability to compete for energy in the world market. Events happen so rapidly in the rich countries and progress comes with such difficulty in the poor that cause seems separated from effect. Jacob Bronowski, in the film made from his book *The Ascent of Man,* made this point. In a dramatic scene, he waded fully clothed into an ash pond at Auschwitz and grasped the mud made of the ashes of millions of Jews and Gentiles. Holding that mud he implored us "to close the distance between the push-button order and the human act. . .we have to touch people." The distance between the push-button order and reality is the distance between the overfed and the starving, the light switch and the coal mine, bureaucraticized war and devastation, the orders and the gas chambers. Energy, like industry, can be applied both to close and to widen these distances. It is a tool, a means, not an end. The labor and environmental revolutions ameliorated some of the worst problems of the industrial revolution. So must a new perspective on energy use. Energy conservation can serve as that extraordinary act necessary to maintain the flow of ordinary events.

P.3. Energy Transitions

A chart of fuel use in the U.S. throughout its history appears as three overlapping waves. It has gone through two major transitions as it moved from solar energy (wood) to coal and then from coal to petroleum. Wood provided the basic fuel until about 1880 when coal first surpassed it. Interestingly, it was not long before the curve of coal use crossed that of wood that oil was discovered (see Figure P.1).

The famous "Colonel" Edwin Drake, who died a pauper, first tapped the oil resource in 1859 in Pennsylvania. It was there, much to the chagrin

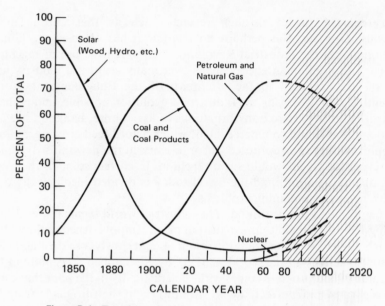

Figure P.1. Transitions in U.S. energy supply. Source: Reference 5.

of those who fumble with energy statistics, that the 42-gallon barrel of oil was introduced, and the world's first oil pipeline was laid. Villages like Pithole and Oil City became wealthy boom towns. Oil which first cost $20 per barrel soon fell to 10 cents. The transformation of the face of the Earth by the automobile was expedited when the first drive-in filling station was opened in Pittsburgh in 1913, not by a Pennsylvania oil company but by one from Texas where oil had been discovered in 1901. As quickly as they had become boom towns, however, Pithole and Oil City became ghost towns, all their economically recoverable oil exhausted. This portent, along with the shortage–glut cycle first evidenced in the price drop from $20 to 10 cents per barrel, presaged the desperation with which the world seeks and uses oil.[2]

Oil overtook coal as America's primary fuel in the1940s at about the same time as nuclear energy was first controlled. Enrico Fermi and 49 others assembled a critical mass of uranium and blocks of graphite moderator under the stands of Stagg Field in Chicago in 1941. Enchanted by prospects of boundless and cheap power, nuclear energy became fixed in the minds of many as the major source of energy, and thus energy projections were drawn in which the curve of nuclear energy use overtook oil sometime in the late twentieth century. This assumption became so widespread by the 1950s that it reigned as *de facto* energy policy for the U.S. for three decades.

With the transition from wood to coal, and later from coal to petroleum, the magnitude of demand for purchased energy changed profoundly. Demand rose from slightly more than two quadrillion BTU (quads) per year in 1859 to more than 20 quads per year in 1941. Now the U.S. consumes almost 80 quads per year, three-fourths of which, is from oil and gas, and some projections of turn-of-the-century energy demand would have us using half again that amount. Such auguries call for policies which are the opposite of what is needed. They ignore basic demographic, economic, and environmental factors which make high energy demand futures implausible if not impossible. They also ignore the fact that energy is a means of supplying amenities, not an end in itself. As we make the transition from petroleum to renewable energy resources, we cannot sustain rapid growth in energy use. This transition must therefore be accompanied by a basic transformation in the way we use energy.

P.4. Conservation Defined

Few concepts are interpreted as diversely as energy conservation. Because world energy prices changed abruptly in 1973 and the simultaneous embargo of oil engendered such sudden responses, most people have come to associate conservation with those curtailment actions that had to be taken quickly to reduce demand. Conservation strategists, however, employ a much broader notion of conservation, one which allies the term with "wise use." Three major strategies are implied by wise use: (1) obtaining higher efficiency in energy production and utilization, (2) accommodating behavior to maximize personal welfare in response to changing prices of competing goods and services, and (3) shifting from less to more plentiful energy resources. All three strategies emphasize technological change that allows smaller energy requirements for a given amenity level. In a real sense, then, energy conservation means substituting ingenuity for energy-intensive living. The main principle guiding the amount of energy conservation desired is the comparison of the real price of energy with alternative goods and services. Conservation is thus viewed by the practitioner as a means of enhancing perceived welfare and as a means of leaving society better off than it otherwise would be.

It follows that waste, too, is an economic term. To fail to make changes which do not affect lifestyle, that is, which do not affect the level of amenity obtained from a given energy-consuming service or product, and which have an acceptable rate of return on investment, is wasteful. Conservation is the sum of those measures which simultaneously save energy and are economically justifiable.

Table P.1. Major End Uses of Energy in the U.S.

End use	Percent of total
Space heating	16
Automobiles	13
Chemicals manufacture	9
Trucks	6
Iron and steel manufacture	4
Lighting	4
Air conditioning	4
Water heating	4
Paper manufacture	4
Subtotal	64
Other	36
Total	100

Source: Reference 3.

Providing the nine basic amenities listed in Table P.1 requires almost two-thirds of the energy used in the U.S. Thus while it might seem that in order to use less energy we will have to sacrifice some of these essentials supplied by energy, such is not the case. The energy required to heat existing buildings, for example, can be cut by more than half with economical weatherization features. Automobiles can be built to obtain 40 miles per gallon at no extra total cost to the consumer or sacrifice in safety and with little loss of comfort and performance. The energy productivity of chemical manufacturing can be improved 20 to 40 percent by the end of the century. Only in emergencies must we curtail our use of energy.[3]

P.5. Energy and Equity

Energy is a scarce resource in an economic sense and the issue of its allocation may be one of the most crucial of our time. Our traditional democratic ideals commit us to an equitable distribution of scarce resources. "That action is best which accomplishes the greatest happiness for the greatest numbers"[4] has become almost an algebraic commandment, but mechanisms for perfectly distributing resources remain as elusive as definitions of happiness.

In market economies, the strategy has been to let the marketplace allocate resources, to let price and preference answer the "Energy for What?" question. Unfortunately, American consumers do not get perfect signals about what energy really costs. Energy users are subsidized by tax credits to energy producers, by road construction appropriations, by energy price regulation, and by other market interventions. Users are also sub-sidized indirectly by costs which are not included in the price of energy. We mean, for instance, the external costs borne by those whose health is af-fected by air pollution, whose homes are damaged by coal mining, or by those miners whose lives are lost in mines built too cheaply, and by inflation and additional national defense costs of oil imports. How to internalize these costs, how to see that the poor have enough of the basic amenities provided by energy, how to assure that energy producers get a fair rate of return on investment, and yet how to maximize individual freedom—these are questions which must be answered by both private and public in-stitutions in developing energy policy and devising conservation strategies. The policy tools available, i.e., price, education, incentives, taxes, regulation, and research and development, to be applied successfully, must be applied with these questions in mind.

P.6. Summary of the Book

What follows is a three-part examination of the prospect for energy demand and conservation. The first part chronicles significant energy demand studies through the 1970s. These studies are analyzed for their assumptions in order to evaluate their results. This analysis includes con-sideration of the factors of energy demand including demographic, economic, and international elements which in the aggregate will determine the course of the U.S.'s energy future. We find that most major energy demand projections are deficient and produce results that call for costly, destructive, and counterproductive energy policies. Part II provides a context of energy supply availability and price. We concentrate on the all-important liquid and gaseous fuels, oil and natural gas, and their potential surrogates. We find that these replacements, whether they be synthetic fuels from oil shale, coal, biomass, solar energy, or electricity from any source, can only cost more than conventional fuels. We note, however, that we could rely upon additional supplies of natural gas as we make our transition to renewable energy sources. The second part of the book also provides a serious look at the health and environmental effects of energy production and consumption. An explanation of the implications of thermodynamics on energy policy is also offered. Parts I and II together serve as a context for the third and final portion, a detailed review of the technical potential for energy conservation. The potential for increasing energy productivity in each of the main energy consuming sectors of the economy, buildings, transportation, and industry, is demonstrated. This potential of course deals with the rational substitution of economically feasible energy con-servation options for the brute force of energy over the period of the next few decades. The Epilogue discusses briefly both the unhappy likelihood that in order to survive an oil embargo we must curtail our consumption of energy drastically in the 1980s and the mechanisms for achieving an effective, equitable degree of curtailment.

The basic conclusions which we reach in The Conservation Revolution include:

1. Influential energy demand studies have seriously overestimated future U.S. energy requirements. Energy consumption in the U.S. presently totals almost 80 quadrillion BTU per year. Projections of U.S. annual energy demand in the year 2000 range from 60 quads to more than twice that amount. It is essential that we ask, "Why do the projections differ so?" To underestimate demand, and thus fail to provide adequate supply, could be economically disastrous; to overestimate demand, prompting serious overinvestment in supply systems such as coal mines, power plants, or synthetic fuels facilities, could be more devastating.

The difference in projections, described in Chapter One, can be attributed to differences in both methodology and working assumptions. The higher estimates generally involved extrapolation of historic trends in ways that are not valid for the longer term, and which ignored major changes in the factors that drive demand. Two common, erroneous assumptions are (1) economic growth and energy use are inextricably related so that energy use must increase as the economy expands, and (2) energy use is price-inelastic to the extent that real price increases will not affect energy use. We now believe, with understanding gained in the use of economic-engineering studies, that over the long term energy productivity can substantially increase through the substitution of capital goods and ingenuity for energy in a way that markedly decreases the energy/Gross National Product (GNP) ratio. The more recent and more detailed studies suggest energy demand levels for the turn of the century of less than 100 quadrillion BTU per year. Given common assumptions. several more of the demand studies we have evaluated would probably support this thesis. The more recent detailed supply studies also tend to affirm the notion that a high energy future simply is not achievable.

2. The basic factors of energy demand are changing, and these changes all point to diminished energy demand growth. Energy use is determined by many demographic, economic, and technological factors including population, labor force, GNP, labor productivity, energy price, and consumer demand for the amenities that energy helps provide. U.S. population growth, a key parameter in almost every resource issue including energy, has slowed considerably and should continue to diminish until the U.S. population levels off sometime in the mid-21st century, perhaps even sooner. The demand for jobs and the level of GNP will be directly affected by the diminution in population growth, and GNP should therefore grow more slowly on an aggregate basis. *Per capita* income will still grow if we can manage to restore historic (pre-1970) improvements in labor productivity. Energy prices, perhaps more than any other single factor, have changed and will continue to change the outlook for energy demand and conservation. From all available evidence, it appears that in real terms energy prices will continue to increase to a level two or more times higher than present. Although price increases have a limited impact on consumption in the short run, over the long run they facilitate economical substitution of more efficient methods of producing a given amenity level.

3. *It is now clear that economic growth is possible without growth in energy demand.* Industrial energy use efficiency has been improving for the last three decades at an annual rate of 1 to 1.5 percent per year despite a real decline in energy prices up to 1973. Energy savings of up to 50 percent in existing households are economically feasible at today's prices. Present average automobile fuel economy could be tripled by the end of the century with less than a 10 percent increase in purchase price and without any increase in total cost of ownership and operation. These kinds of cost-effective opportunities should enable the long-term downward trend in the ratio of energy use to GNP to continue for decades.

4. *The price of energy can only go higher.* Petroleum in the U.S. is discovered in increasingly smaller quantities per deposit and in harsher environments. The costs of exploration and production are thus increasing, as we discuss in Chapter Four. The demand for oil on the world market further escalates petroleum prices as developing countries needing oil for furnishing basic amenities compete with rich nations. Either demand for development in the Third World will come to a catastrophic halt, or the demand for oil, and therefore the upward pressure on the price of oil, will continue.

Solar energy will ultimately place a lid on energy costs because solar energy is renewable. But solar energy could cost two to three times as much as the average price of energy today. A nuclear breeder reactor would also supply energy (electricity) but at a cost of perhaps three to four times that of today's oil and natural gas. Use of coal in an acceptable way will be possible, but both costly and difficult, and will be complicated by such problems as the carbon dioxide greenhouse effect for which there seems little promise of a technological fix. Fusion is no more than a gleam in the eye of optimists. No energy supply technology that we know of can reverse the rising cost of energy. Large quantities of natural gas are available, however, at prices between its current regulated price and the cost of synthetic substitutes or of electric energy.

5. *U.S. oil consumption hampers Third World development, is a threat to world peace, and shortens the time over which energy supply transitions must occur.* The U.S. not only consumes 30 percent of all the oil produced

in the world, but it also is a slight net importer of energy embodied in goods. This means that U.S. energy consumption is not recycled to other nations in the form of food or manufactured products. This situation threatens peace not only by creating tension in the oil-producing nations, but by widening the gap between the haves and the have-nots.

The income gap between the developed and the underdeveloped nations is 12 to one. That is, the annual income in the wealthiest nations is 12 times as great as that in most poor nations. Such disparity takes its toll in human welfare and in the ability of nations to live in peace. Chapter Three describes the disparity in energy consumption not only between the U.S. and the poor countries, but also between the U.S. and other developed countries.

Oil or a close substitute is a prerequisite for development. It is an essential ingredient of fertilizers, for freight transportation, for much of the infrastructure necessary for economic growth. To the extent that excessive U.S. demand drives the poor countries of the world out of the oil market, it forecloses on the future of millions.

6. Energy price controls harm most the poor, induce unnecessary environmental damage, forfeit national security, and ultimately lead to higher energy costs. Energy price controls subsidize middle and upper income consumers more than the poor simply because the poor buy less energy. Natural gas, for example, became unavailable under price controls forcing consumers to use vastly more expensive electricity. Price controls stimulate demand which leads to scarcer and ultimately more expensive fuels, and, as a result, higher inflation which most harms lower income citizens. European consumers pay a higher price for energy, but a smaller percentage of their incomes for energy because they have taken the available opportunities for energy conservation.

To the extent that energy conservation opportunities are lost because consumers do not receive correct signals about the real cost of energy as a result of energy price controls, additional amounts of energy must be produced and converted, causing pollution. The result is the unnecessary death and suffering of, for example, coal miners and inhabitants of areas with poor air quality. Frequently, the victims of the deleterious effects of energy production and consumption are those least able financially to protect themselves by moving or to obtain adequate health care once afflicted.

Regulated oil prices have helped stimulate gasoline consumption to a volume that has eroded our national security. A combination of taxes and enlightened pricing policy has driven European gasoline prices to a level two to three times that in the United States and has helped drive per capita demand for gasoline to one-fourth that of the U.S. Similar policies in the U.S. would help extricate us from dependence on Middle East oil.

7. *Environmental protection policies should not be relaxed to foster energy production. Most are cost-effective.* Federal environmental protection statutes are frequently fashioned at levels of control that offer less than the economically optimum protection of public health and natural resources. The Coal Surface Mining Control and Reclamation Act, for example, added only $0.02 to $0.10 to the cost of a million BTU of electricity, which averages more than $11.00 per million BTU, but can prevent damages as high as $4000 per household in strip-mined watersheds. The use of Flue Gas Desulfurization (FGD) or scrubbers to remove the acid-rain-causing sulfur from power plant smoke may add $0.50 to $1.00 per million BTU, but could save or extend thousands of lives which carry value we do not attempt to calculate. Catalytic converters on automobiles cost trivial amounts of energy but also can save thousands of lives. To avert attaching the total cost of energy use to the price the consumer pays would be a false bargain. Chapter Six amply illustrates that while environmental protection is not a luxury, it is generally a good bargain. Since energy use is a major source of environmental insult, and since envirionmental protection as currently practiced is usually cost-effective, we should not relax our efforts to control the effluents of energy production. To the contrary, we may wish to strengthen these efforts and make them permanent. This is a problem that private enterprise, by definition, cannot solve, and that individual states are often ill-equipped to handle, since pollution, and the energy markets which create it, is not constrained to stay within state lines. The federal government should and must continue to exercise its responsibility to protect the health and welfare of its citizens by controlling environmental pollution.

8. *Energy conservation methods could reduce year 2010 energy demand from a projected and conceivable level of more than 120 quads to as little as 60 quads, 20 quads of energy less than is used today.* New nuclear or coal fired electrical generating plants will produce electricity for $15.00 per million BTU (1980$). Ceiling insulation in an uninsulated house will save energy at $1.00 per million BTU, while wall insulation and storm windows will save energy at a cost of $2.00 and $3.00 per million BTU, respectively.

Synthetic fuels plants will produce gasoline from coal or oil shale to retail at $15.00 to $30.00 per million BTU, compared with 1980 costs of $10.00 per million BTU for gasoline. Synfuel plants could, at most, produce about two million barrels of oil-equivalent per day by 1995 (see Chapter Five). By 1995, cars obtaining 40 miles per gallon could save two million barrels of oil per day at virtually no extra cost to the consumer (see Chapter Nine). If such efficiency improvements were assured, we might not find markets for expensive synthetic fuels.

Depending on the specific source, synthetic natural gas for industry will cost $8.00 to $20.00 per million BTU. But industrial conservation devices could save many quads of energy each year by the year 2000 for less than industry pays for oil and natural gas today (see Chapters Five and Ten).

Reasonable energy demand estimates for the U.S. range from 60 to 120 quads per year in the early part of the next century. On the average, every quad of energy that could be conserved but is not will represent a loss to the economy in excess of $10 billion. The opportunity cost of a failure to invest aggressively in energy conservation will be staggering.

9. Achieving the potential for energy conservation in buildings will require policy initiatives on the part of private utilities and investors as well as federal, state, and local governments. Federal tax incentives and grants for buildings sector energy conservation currently average about $3 billion per year. The 15 percent residential tax credits offered by the federal government to homeowners as an incentive to invest in conservation, and the federal grants for weatherizing schools, hospitals, and low-income families' homes are thus helpful. But the amount of financing presently needed to capture most cost-effective energy conservation opportunities in the buildings sector may run as high as $100 billion. Thus, even when private investment in conservation is considered, at the present rate of investment decades will be required to retrofit buildings to a degree needed today.

Clearly, the private sector must undertake this enormous job if it is to be accomplished as needed within the next ten years. Electric and gas utilities should find it in their interest to provide financing for conservation in their customers' homes. Utilities' customers could share the benefits of energy saved for a cost far below the cost of new energy supplies. Regulated utilities should be given incentives to make such investments; public utility commissions should consider requiring these investments. Utilities are unique in having the financial capabilities, the expertise, and the incentive to invest in conservation in the buildings sector. Financial institutions should find conservation investments far more secure than risky energy production ventures and should find utilities to be a convenient vehicle for secure investments.

Utility arrangements for retrofitting rental housing and commercial buildings should be developed and encouraged. Renters are caught in the dilemma of having no means of investing in weatherizing their buildings, often having landlords who do not pay the utility bills and therefore have little incentive to insulate, caulk, and weatherstrip rental units. Commercial building owners and tenants frequently have too little knowledge or too little capital to realize potential conservation savings. Utilities could provide energy saving services as easily as energy services and recover their costs on the monthly utility bills.

Performance standards for new buildings should be developed and imposed. Buildings built today are constructed at a fraction of optimum thermal integrity. The market clearly fails to require optimum conservation investments when builders determine the level of weatherization that goes into a new building and the utility bills are paid by a new homebuyer who chooses a home based on location, style, price, etc. more than on thermal integrity. Appliance efficiency standards are meritorious for similar reasons. Chapter Eight explores these and other buildings energy conservation possibilities.

10. The transportation sector, because it consumes by far the largest amount of oil, should be our highest conservation priority. A standard of 40 miles per gallon for early 1990s model cars, and 30 miles per gallon for light trucks and recreational vehicles, would yield major petroleum savings without driving up the total cost of operating a vehicle. Federal assistance to automakers to achieve and exceed this standard are in order. Tax disincentives for long distance trucking of freight would encourage a further shift to rail and barge. Incentives to increase the load factors of all vehicles, especially to encourage car and van pools, are needed. A substantial tax on gasoline and diesel fuel would effect curtailment of gasoline consumption in the short term and could help us survive the years between now and when conservation begins to extricate us from dependence on foreign oil. Such taxes could only be acceptable if rebated, however, or used to replace regressive taxes, such as state sales taxes on food. The Epilogue touches on this critical issue.

11. Industrial energy conservation requires institutional as well as technical innovation. Industrial energy consumption per unit of output has declined at a rate of 1 to 1.5 percent per year for the past four decades. Industrial output can continue to grow substantially without increasing total energy demand. Energy management and retrofits of existing process equipment, as well as completely new manufacturing processes, will facilitate this task, as we illustrate in Chapter Ten. Industrial cogeneration, in addition, could furnish some 100,000 megawatts of generating capacity. But these possibilities will not be realized without major policy changes.

Energy efficiency standards for industrial equipment could be imposed but would probably be a nightmare to enforce. Incentives such as rapid depreciation for investment in energy-saving equipment probably are a more efficacious alternative. Incentives and/or regulations are needed despite the fact that industry should, in its own self-interest, minimize energy costs in an optimum relationship with the other factors of production such as labor and capital. Competing demands for industrial capital often dictate that investments such as ones designed to increase a corporation's share of a particular market be given top priority. Such opportunities may come only once, and whereas conservation opportunities do not often go away, these can be captured at leisure. Thus, incentives such as rapid tax write-offs should be fashioned to accelerate industrial investment in energy conservation.

Exciting new institutional arrangements may create prodigious energy savings. Utility industrial partnerships can be created to share the energy and capital costs of generating steam for generating electricity and processing industrial goods, or third party corporations can be developed to fill a vacuum left in this role by uninterested utilities and industry. The fuel required to "cogenerate" electricity in this way is roughly half that of central station power plants. Industries can save as much as $1.00 to $2.00 per million BTU of steam, and utilities can generate power at two-thirds the cost of that produced in a new power plant. Federal constraints on the use of natural gas and low-grade oil should be removed for cogeneration. Steam sales, regulated as a public utility in 24 states, should be deregulated.

Energy management corporations, companies that would serve the function of capturing energy conservation opportunities by earning a percentage of energy bills saved, could deliver important energy savings to both industry and commercial consumers, and possibly homeowners as well. They should experience rapid growth in the U.S., and regulatory constraints on their effectiveness should be removed.

12. Growth in demand for electrical generating capacity will continue to be far less than historical. Despite the need to switch from petroleum to electricity for many end uses, electrical capacity is presently overbuilt in most parts of the nation and a surfeit of kilowatts may exist until late in the 1980s, depending on public policy and industry actions that remain pending with respect to new plants, especially nuclear. There are areas such as California where the reserve margin is growing thin due to rapid demand growth or lack of supply expansion, but that situation is not typical in the U.S. The present overcapacity is due to excessive power plant construction in the face of rapidly slowing demand growth. It is this overcapacity of very expensive plants that is chiefly responsible for skyrocketing electric rates.

Additional electrical capacity when needed could be supplied to a considerable extent by industrial cogeneration (see Chapter Ten) and by utility load management. Cogeneration plants can be built with much less lead time and with far more flexibility in following utility loads. The existence of a large cogeneration potential is good insurance that adequate electrical supply will be available in the event that electric growth again accelerates. The high cost of electricity and the near saturation of many electricity-intensive appliances such as air-conditioners should continue to inhibit electrical demand growth, however.

13. The potential for solar energy is large, but. . . . Solar water heating is cost-effective in most areas of the country today, and water heating annually requires more than three quadrillion BTU in the U.S. Biomass, as we indicate in Chapter Five, could supply ten quads or even more by the end of this century.

But even renewable energy must be conserved. Diffuse sources of energy can never be hoped to satisfy U.S. energy needs unless energy conservation opportunities are taken. Moreover, solar energy sources will be expensive, and although we will enjoy solar energy's benign attributes, we will have to carefully apply it in most cases in order to make it affordable. Thus, energy productivity must become a first priority.

Aggressive solar energy research and development could carry large benefits, especially in the area of direct conversion of sunlight to electricity and in the conversion of biomass into liquid fuels.

14. The 1980s will be difficult. Throughout this book we emphasize that the energy system can only change slowly without major economic and social penalties. While we identify plausible ways to resolve our energy problems over a period of several decades, we have spent little time on the very likely event of short-term disruptions. We can, however, draw from our work on longer time horizons some relevant observations about changes that can occur in less than five years:

(a) Domestic natural gas and, to a lesser extent, oil production can be spurred, particularly by price deregulation.
(b) Massive amounts of energy can be saved by retrofitting residential and commercial buildings, and by investments in industry.
(c) The transportation infrastructure can be made substantially more efficient by modal shifts and vehicle occupancy improvements while we replace existing capital stock with more efficient equipment.
(d) Strategic reserves of oil and gas can be built during times of excess international supply.

Perhaps most importantly, workable emergency curtailment plans can be made. We return to this critical but largely and unbelievably neglected issue in the Epilogue.

. . . Man is turned into a chicken or a rat, ruled over by an elite that derives its power from the wise counsel of intellectual aides who actually think that men in think tanks are thinkers and that computers can think; the counselors may turn out to be incredibly insidious and instead of pursuing human objectives, may pursue completely abstract problems that had been transformed in some unforeseen manner in the artificial brain.
—HANNAH ARENDT
"On Violence" from *Crises of the Republic* (1970)

Dreams and predictions ought to serve but for winter talk by the fireside.
—FRANCIS BACON
"Of Prophecies" *(Essays,* 1625)

Part One

A Short History of the Future

Introduction to Part One

Francis Bacon would have had us relegate predictions "to. . .winter talk by the fireside." We cannot, unfortunately. The exigency of energy will not allow it. We know now that a five billion dollar synthetic fuels plant may take ten years to build and may be quite a risk to public health. Society cannot afford to spend billions or risk catastrophe on projects it does not need. Bacon notwithstanding, good estimates of future energy demand are essential.

Chapter One, "A Chronicle of Forecasts," examines several influential U.S. energy demand projections. It begins with a short introduction to "scenario analysis," then proceeds to detail the results of some of the more interesting energy demand studies. One interesting thing about these studies is that they differ drastically. Some, ignoring the effect of energy price on the potential for conservation, estimate that energy

17

demand in the U.S. will be 250 percent greater by the end of this century than at present. More recent studies have described the feasibility of a relatively low energy future. Another major point of interest is the difference in the estimates of electrical demand. The projections of the number of additional large power plants (1000 megawatts each) needed for the future range from a few tens to more than 1000.

Chapter Two, "The Elements of Energy Demand," describes in detail the economic and demographic forces that "drive," or create, energy demand. Population, GNP, labor force and productivity, energy price elasticities, and other ingredients of energy demand are assessed, primarily for how the demand studies examined in Chapter One compare in their basic assumptions. We offer some judgments on the validity of these assumptions, and, by inference, on the merit of the various demand projections.

Chapter Three expands the analysis begun in Chapters One and Two into the international arena. Specifically, it examines the prospects for closing the income gap between the rich and poor nations insofar as energy use is concerned. Chapter Three also compares the energy productivity performance of the U.S. with that of other industrialized nations and makes observations on the importance this performance has for the well-being, if not the survival, of the rest of the world.

Table 1.1 lists and briefly describes each of the major demand studies evaluated. Figure 1.1 summarizes the results of each study, and Table 2.1 summarizes the most important assumptions made in each. Figure 1.1 graphically compares their energy demand projections.

Prophecy is the most gratuitous form of error.

—GEORGE ELIOT
Middlemarch (1871–1872)

Chapter One

A Chronicle of Forecasts

1.1. Scenario Analysis

Certain methods of divination enjoyed wide currency in the ancient world; for example, events were foretold by examining the entrails of sacrificed animals or humans, by drawing pebbles from a heap, by numbers, by dots made at random on paper,* and, of course, by interpreting oracles. Most methods of divination, however, have disappeared in the modern world. Yet because modern society requires extensive planning, auguring is now even more essential. And because energy is such an essential component of modern life, few fields of late have experienced more oracles generating numbers and dots on paper.

Divination, however, implies actually foretelling events. It is important

---

*Known as geomancy.

to distinguish between forecasts and projections because the former are tenuous in the extreme. As one observer put it, "having been exposed to the nightly miscalculations of the weather forecasters, the American public deserves to be a little skeptical about somewhat more complex efforts to predict the future."[1] While some analysts may have believed their work to be forecasts, the general approach is to project plausible events by using the best available tools and by making clear to the user the assumptions employed. In this way, scenarios of possible futures are written.

Scenario analysis is at least as old as literature, though it was popularized in the eighteenth century by figures like Walter Petty, who wrote about the dangers of underpopulation. Petty wanted to move Ireland's entire population to England to concentrate the populace, and he was opposed to all colonial settlements because he thought they were a drain on population. A somewhat wiser Thomas Jefferson wrote population scenarios for the U.S., such as the following[2]:

> Should this rate of increase continue, we shall have between six and seven million inhabitants within 95 years. If we suppose our country to be bounded, at some future day, by the meridian of the mouth of the Great Kanhaway (within which it has been before conjectured, are 61,491 square miles), there will then be 100 inhabitants for every square mile, which is nearly the state of the population in British Islands. (pp. 162-163)

Note the importance of the rate of increase, the assumptions required to generate the projection, and the projection itself. This trend extrapolation (for the rate), good guesswork (for total land area), and some simple calculations were combined to compose Jefferson's scenario. Thomas Malthus' famous forecast of cataclysmic overpopulation assumed that population growth would continue unabated except for countervailing events of pestilence, starvation, and war. His assumption has been proven wrong, of course, in almost all developed countries, because he failed to anticipate the effects of technology, including birth control.

1.2. Major Energy Demand Scenarios

The decade of the 1970s was replete with energy demand studies. From among them we have chosen eight for examination in the following chapter. These eight studies, or their close relatives, went a long way toward shaping America's thinking about energy, and show clearly the reason why the nation's thinking about the subject is so diverse. Estimates, for example, of total demand for energy for the year 2000 ranged from 56 to 200 quads per year, but current use is roughly 80 quads per year. Estimates of demand for electrical energy for the year 2000 ranged from 23 quads to 75 quads, a difference of approximately 1000 power plants. Estimates of our dependence on foreign oil ranged from zero to 21 million barrels per day, two and one-half times our current quantity of imports. Little wonder then that the general public suffers from confusion about energy policy. It not only extends to, but also derives from, the experts.

But a consensus of sorts has been forming. Viewed chronologically, the estimates of future energy demand have been getting smaller. Smaller estimates result partly from refined analytical techniques, partly from the drastic change in prices over the last decade, and partly from more comprehensive treatment of the issue. Whatever the case, our ability to understand the reasons which underlie projections of demand will affect our ability to plan and control our energy future.

The major energy demand studies we consider in detail are listed and compared in Table 1.1 (see also Figure 1.1). Other energy demand studies of an international nature will be discussed in Chapter Three. Certain U.S. studies such as Project Independence were not included for reasons of economy of detail—the quantity of information contained in just one study is overwhelming. The methodology and results of the Project Independence work, however, closely approximate other analyses which are included, and these similarities are noted. Please note that when several scenarios were generated by one analysis, the one(s) most representative of that work was chosen for comparison. Generally, we have relied on the recommendations of the authors of the eight major studies in this choosing.*

1.3. Discussion of the Demand Scenarios

A detailed discussion of each of eight major U.S. energy demand studies and certain representative scenarios generated by each follows. This discussion provides background to the energy policy debates of the 1970s and prepares us to explore the ingredients of energy demand in the next chapter.

1.3.1. Guidelines for Growth of the Electric Power Industry (FPC-1970)

The initials FPC-1970 (Federal Power Commission, 1970) denote a classic study of energy demand, although only electrical energy was

---

*The authors of this book would like to acknowledge their participation in two of the studies reviewed: Gibbons, as chairman of the Demand/Conservation Panel of the National Research Council's Committee on Nuclear and Alternative Energy Strategies, and Chandler as an author of *Economic and Environmental Impacts of a U.S. Nuclear Moratorium 1985–2010,* published in 1976 by the Institute for Energy Analysis (second edition published in 1978 by MIT Press).

Table 1.1. Major U.S. Energy Demand Scenarios (1980 Energy Demand ≈ 78 Quadrillion BTU)

Abbreviation	Title	Author	Year projected	Energy demand in year projected (quadrillion BTU per year)
FPC–1970	Guidelines for Growth of the Electric Power Industry	U.S. Federal Power Commission	1990	140
Limits to Growth–1972	Limits to Growth	D. Meadows et al., for the Club of Rome	2000	165 to 300
AEC–1973	The Nation's Energy Future	U.S. Atomic Energy Commission	2000	150 to 200
EPP–1974	A Time to Choose: America's Energy Future	Energy Policy Project of the Ford Foundation	2000	188 (Historical growth) 124 (Technical fix) 100 (Zero energy growth)
USBM–1975	United States Energy Through the Year 2000	U.S. Bureau of Mines	2000	163
FEA–1976	The National Energy Outlook — 1976	U.S. Federal Energy Administration	1985	93 (Conservation) 100 (Electrification)
iea–1978	Economic and Environmental Impacts of a U.S. Nuclear Moratorium, 1985–2010	Institute for Energy Analysis	2000	101 (Low)
CONAES–1979	Alternative Energy Demand Futures to 2010	National Academy of Sciences	2010	58 ("A*") 73 ("A") 94 ("B") 133 ("B'")

Source: References 3, 4, 5, 7, 8, 9, 11, 12.

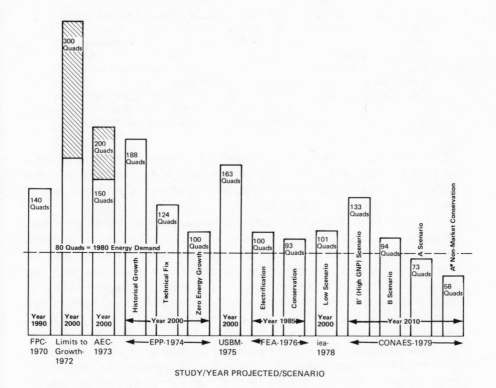

Figure 1.1. Major U.S. energy demand scenarios.

seriously evaluated. *Guidelines for Growth of the Electric Power Industry*[3]
attempted to be an econometric study of electrical energy requirements
through the year 1990. The study appeared amidst the growing awareness of
the environmental movement of the 1960s. The Atomic Energy Commission
had not long before lost the historic *Calvert Cliffs* case and was thereby
required under the National Environmental Policy Act (NEPA) to prepare a
full environmental impact analysis for each new nuclear power plant. On
the environment and the need for new sources of electric power,
FPC-1970[3] had the following to say:

> Those who are concerned about the environmental impact of electric power
> operations, and who would seek to curtail the growth of electrical demand on this
> account, sometimes lose sight of the fact that our society has an indoor as well as
> outdoor environment and that the two are interdependent. We cannot expect to
> heat, light, refrigerate and power our indoor environment on a massive scale, as
> our society has become accustomed to doing, without affecting the outdoor en-
> vironment. Conversely, restraints placed on allowable changes in outdoor con-

ditions must sooner or later affect indoor standards. Clearly what is needed is a balanced approach with all relevant factors properly weighed. Clearly, also, a situation of imbalance exists today in the nation's overall environmental ledger. Too much attention has been paid in the past to our indoor standard of living with too little heed to the side effects on the quality of the natural environment. Fortunately there is growing recognition that this problem exists and, as was mentioned earlier, a crusade to improve and protect the natural environment has gathered momentum. However, all pendulums swing back as well as forth and so there is need to guard against the possibility of overcorrecting and thereby creating a new imbalance. (pp. I-1 to I-13)

These words were written before passage of the Clean Air Act Amendments of 1970,* which greatly affected energy use and production, and of course before the strengthening of the Federal Water Pollution Control Act in 1972 (now called the Clean Water Act).

In the FPC-1970 study, three different macroeconomic modeling efforts were made. One was based on two levels of income and two electric-to-gas price ratios for an estimated population level. One substituted the number and size of households for the population parameter, while the third substituted GNP for income and combined it with the residential demographic variables. The third effort was basically an extrapolation of trends. Because the FPC had so accurately estimated the 1970 level of energy demand (their 1964 modeling effort had misjudged 1970 demand by only three percent), the authors were willing to call their work a "forecast."

The FPC predicted that total energy growth would average 3.4 percent per year until 1990 and that electrical demand growth would average 6.7 percent. Total energy use would therefore rise from 70 quads per year to 140, a doubling in only twenty years. These figures were in accord with historic data which have shown a doubling in energy consumption every twenty years in the U.S. since 1850, but the unforeseen energy price increases and the recession experienced since the study have effectively devastated its conclusion. Energy growth in the decade of the 1970s averaged less than one percent per year.

1.3.2. Limits to Growth (1972)

Limits to Growth,[4] commissioned by the Club of Rome, perhaps more than any "official" (i.e., government-sponsored) document, brought the growth debate to the public's attention. The study dealt with several critical

---

*The Clean Air Act was first passed in 1963, although it was not until the 1970 Clean Air Act Amendments were passed that the legislation made a serious impact. The Act was amended again in 1976. The amendments of 1970 and 1976 are frequently referred to as "The Clean Air Act."

natural resources, and energy was, of course, central to that analysis. A computerized extrapolation of current consumption trends described a disastrous disjuncture of the growth curves of population and food supply and of resource use and pollution. Several scenarios (including one in which the authors doubled the computer inputs for the availability of key resources, despite their belief that such quantities do not actually exist) climaxed in what was called a collapse mode, the sudden and drastic decline in the human population as a consequence of famine or environmental poisoning. The report was aprocryphal. *Limits to Growth* quite correctly made the point that exponential growth in finite systems is unsustainable, but its analysis was based on an aggregate, or macroeconomic model. While such models are useful in describing economic behavior over short periods of time and help policymakers deal with inflation, unemployment, and related problems, they generally fail to consider responses to price. One such failure was to substitute conservation for energy to provide a given level of output. The authors of *Limits to Growth* also failed to take into account advances in the technology of pollution control. However, microeconomic/engineering methods, which better account for the effects of price and technological changes, have only recently been applied to projecting energy futures.

    Limits to Growth was published just in time for the Arab oil embargo in 1973. Oil prices quadrupled, thus reversing the 20-year trend in declining energy prices. Much of the world's machinery was made obsolete overnight simply because it had been built to operate on two-dollar-per-barrel oil. With oil now more than 30 dollars per barrel rather than two dollars, it has become obvious that many more conservation practices and much more major capital investment in conservation are economical. The authors of *Limits to Growth* suggested that "the political question may arise long before the ultimate economic one" (p. 75). But with or without cartel, oil will not simply be used up at a certain price and then suddenly be exhausted. On the contrary, scarcity should gradually drive up the price and extend the supply, thus allowing time for adjustment, political embargoes notwithstanding.

1.3.3. The Nation's Energy Future (AEC-1973)

    Because the U.S. Atomic Energy Commission's (AEC) mission included the development of nuclear power, it had for some time prepared electrical energy demand forecasts. Nuclear enthusiasts are said to have forecast in the 1950s that nuclear power would make electricity so cheap that it would not need to be metered. Its efforts to promote the peaceful use of the atom around the world were irreverently tagged by some skeptics with the term "Kilowatts for Hottentots." Through a public information program, the AEC distributed widely the notion that the demand for energy would equal or exceed historical rates. Since a high demand growth rate would rapidly advance the inevitable depletion of conventional fuels, the AEC argued, the rapid development of nuclear power, especially a breeder reactor, was made essential.

The Nation's Energy Future,[5] an energy research, development, and demonstration report to the President in 1973, was actually the predecessor of *Project Independence,* which was published in 1974 by the Federal Energy Administration (FEA). Like its more famous offspring *Project Independence, The Nation's Energy Future* examined ways to increase the domestic supply of energy in order to reduce or eliminate oil imports. The study took as its baseline scenario of demand one which differed little from the common wisdom of the time, a forecast from "Understanding the National Energy Dilemma,"[6] which predicted year 2000 demand in the 150 to 200 quadrillion BTU-per-year range.

The Nation's Energy Future, however, marked a great break with conventional wisdom. It forecast a 1980 demand level of 90 quads without conservation, and only 80 quads with conservation. The AEC report, compiled under the direction of Dixie Lee Ray,* with its remarkably accurate estimate, contained an error, however. It forecast our independence from imported oil by 1980, an error influenced no doubt by then-President Nixon, who was bent on achieving total oil independence in a short enough time to make it a political achievement during his administration. U.S. oil imports in the early 1980s will total about 12 quads per year, or 5 to 7 million barrels per day, provided willing suppliers can be found.

1.3.4. A Time to Choose (EPP-1974)

A Time to Choose: America's Energy Future[7] was commissioned by the Ford Foundation in 1970. The Energy Policy Project (EPP), headed by S. David Freeman (subsequently appointed Chairman of the Board of Directors, The Tennessee Valley Authority), produced about twenty volumes on energy issues ranging from conservation in industry to coal surface mining in the western U.S. *A Time to Choose* soon became controversial. The cause was "Zero Energy Growth," one of its three scenarios of future energy use. ZEG was mistakenly named not only because it polarized the issue, but also because in the ZEG scenario, growth in energy

---

*Chairman of the Atomic Energy Commission at the time of publication of *The Nation's Energy Future.*

was not really zero. Growth would, according to the illogically named scenario, continue until 1985 at a rate which today might actually be considered high.

EPP's two remaining scenarios were "Historical Growth" and "Technical Fix." "Historical Growth" was basically a trend extrapolation intended as a baseline against which the engineering and economic studies could be compared. In the report's words:

> The "Historical Growth" scenario examines the consequences of continuing growth in energy consumption for the remainder of the century at the 1950-1970 average rate of 3.4 percent per year. Industry and government planners, more afraid of being blamed for energy shortages than for energy surpluses and waste, typically plan supply expansion based on a continuation of past trends. They assume that demand will materialize, stimulated if necessary, through advertising, subsidies, and promotional pricing. Accordingly, we selected the 3.4 percent figure for analysis because it is in line with many recent government and industry forecasts. (p. 20)

The "Historical Growth" demand forecast 116 and 188 quads in 1985 and 2000 respectively, not very different from the consensus of the time.

The "Technical Fix" scenario, however, put the 1985 total consumption at 91 quads. This projection was generated by using a microeconometric model (the Hudson–Jorgenson model), coupled with analysis of the engineering potential for conservation as a response to increased energy price and government policies. This scenario was intended to explore the potential for significant reductions in the growth rate of energy consumption without seriously affecting improvement in the standard of living between now and the end of the century. Two criteria were applied to potential conservation measures before they were admitted into the scenario: (1) that a significant energy savings should be possible with existing technology, and (2) that the savings should be achieved in a way that would save the consumer money. Given the energy price increases of the 1970s, these criteria admit a large number of such measures.

The "Zero Energy Growth" scenario clearly was meant to show that energy growth could be decoupled from economic growth. On this point the authors of EPP were very specific:

> Our own research confirms that it appears feasible to achieve zero energy growth after 1985, while economic growth continues at much the same pace as in the higher energy growth scenarios. The mix of the economy would of course be different.
> . . .there would be a greater emphasis on services—education, health care, day care, cultural activities, urban amenities such as parks—which generally require much less energy per dollar than heavy industrial activities or primary metal processing, whose growth would be deemphasized. . .this does not mean that people would lack the valued material amenities of the higher energy growth scenarios. Rather. . .there would be a premium placed on durability and quality of

consumer goods, so that production each year could be lower. Also, material
substitutions would be encouraged. As a prime example, throwaway cans and
bottles would be discouraged in favor of reusable containers. (p. 82)

The scenario called for a leveling off in energy demand at about 100
quads per year shortly after 1985. In addition to those economically feasible
measures taken by 1985, reduced demand would be effected by means of
energy consumption taxes and other constraints on use. Taxes would increase the price of fuels so that substitution of energy saving capital investments could be made economical while simultaneously encouraging
shifts to (less energy intensive) services. For all its notoriety in 1974, the
"Zero Energy Growth" scenario now seems rather tame.

1.3.5. Energy Through the Year 2000 (USBM-1975)

United States Energy Through the Year 2000,[8] was an attempt at
forecasting by the U.S. Bureau of Mines (USBM). First published in 1972
and revised in 1975, this short but influential forecast was basically a trend
extrapolation. Its authors explained:

> This forecast of future energy consumption and supply is based on the
> assumption that existing patterns of resource utilization will continue (p. 2). . . . The
> potentials for conservation were not explicitly factored into the base forecast. The
> role of conservation in reducing future energy consumption is not well de-
> fined. . . . Complicating the picture is the nonadditive nature of the potential
> saving. This should not be interpreted to mean that energy conservation was
> neglected. Implicit in the declining energy/value added for the industrial sector is
> an assumption that industry will become more efficient. The same is true of the
> declining heat rate for electrical power plants. It should be noted that these will
> probably represent more a response to energy price increases than to governmental
> action. (p. 26)

The U.S. consumed about 71 quads in 1970. The USBM, in 1972,
forecast a consumption total of 192 quads for the year 2000. In the 1975
revision it forecast a consumption level of 163 quads for the year 2000. In
that year, 75 quads, an amount greater than the total consumption of 1975,
would go to electrical generation alone. As for total energy demand, even
the revised (1975) estimate would have us using twice as much energy by the
end of the century than we use now.

1.3.6. The National Energy Outlook (FEA-1976 and 1977)

The National Energy Outlook—1976[9] was the successor to, and
essentially a revision of, the Project Independence study (1974), and
produced a "Business As Usual" (BAU) scenario. In addition seven other

projections each to 1980, 1985, and 1990 were performed using the Project Independence Evaluation System (PIES). The PIES model was composed of three major components: a macroeconometric demand model, a series of regional (U.S.) energy supply models, and an integrating model. Energy demand in the PIES model is driven largely by the growth rate of the Gross National Product (GNP), which is an assumed output variable in the first model. Demand is moderated, however, by consumers' response to fuel price changes. Fuel prices are determined by the supply models. Despite the sophisticated structure of the model, the results are driven primarily by exogenous variables such as GNP and price elasticities.

The results were predictable. The BAU scenario has the U.S. using almost 100 quads by 1985, but it assumes, among other things, that the price of oil does not increase above $13 per barrel.

A "Conservation Scenario" was also generated which incorporated a full set of demand conservation initiatives, including increasing the efficiency of autos, the use of van pools, improving thermal efficiency of buildings, accelerating industrial conservation, improving airline load factors, and eliminating gas pilot lights. The result is that total consumption in 1985 is reduced from the BAU level of almost 100 quads to 93.

In order to demonstrate a switch in the economy from oil and gas, an "Electrification Scenario" was included. It showed the impact of rapid expansion in electrical generation and indicated that an end-use level of electrical supply of 1.1 quads per year higher than in the "Business As Usual" projection could be managed by 1985. Total consumption in this forecast would slightly exceed 100 quads per year.

All FEA-1976 scenarios forecast energy self-sufficiency for the U.S. by 1985. It is significant that this study's successor, the *1977 National Energy Outlook,*[10] projected demand for the year 1985 to be lower than the 1976 projection by almost 9 percent in the Reference (or "Business As Usual") scenario. The 1977 study concluded that a two percent annual increase in imported oil prices would decrease 1985 consumption by one quad, and that "if the real GNP growth rate were to fall to 3.8 percent between 1975 and 1980, and to 2.3 percent between 1980 and 1985 compared to the 9.8 percent and 3.3 percent respectively projected in the Reference Case, *projected consumption would drop by 8 quads per year by 1985"* (emphasis added).

1.3.7. Economic and Environmental Impacts of a U.S. Nuclear Moratorium (iea-1978)

The first major effort since the Energy Policy Project that envisioned a relatively low energy future was a study called *Economic and En-*

vironmental Impacts of a U.S. Nuclear Moratorium, 1985–2010.[11] The study was funded by the National Research Council and the Energy Research and Development Administration and was performed by the Institute for Energy Analysis (which we will designate iea in order to distinguish it from the International Energy Agency, IEA) in Oak Ridge, Tennessee.

The main thrust of the study's eight volumes was directed at the implications of a U.S. nuclear moratorium in which no new nuclear power plant construction would be permitted after 1980 (although continued operation of all nuclear plants in operation by 1985 would be allowed). Of more significance, however, may be the results of the energy demand analysis which was generated as a context for the moratorium issue. A "High Case Scenario" projected demand at 126 quads by 2000, while a "Low Case Scenario," which the authors claimed to be more credible, called for "only" 101 quads. Corresponding "Low Case" projections put 1985 demand at 82 quads, 2010 demand at 118 quads. The "High Case" foresaw a total U.S. demand level of 159 quads in the year 2010. Because demand was relatively low with and without a nuclear moratorium, the economic and environmental impacts of such a moratorium were judged to be smaller than might have been expected.

Regarding their methodology, iea had this to say:

> Our projections of demand are in no sense "normative"—that is, we do not suggest what the energy demand, and by implication the life style, ought to be. Instead, our projections result from fairly straight-forward extrapolations of historic trends that determine energy demand. In this sense, we would describe our projections as "surprise free." Indeed, although the aggregate energy demands and GNP increase rather modestly, the energy demands per capita and GNP per capita increase at rates comparable to or higher than historic rates. (p. xxviii)

It at first seems contradictory that trend extrapolations would produce a scenario of energy demand *lower* than the "Historical Growth" or "Business As Usual" scenarios of EPP and FEA, or indeed, lower than the FEA's "Conservation Scenario," but approximately equal to the EPP "Zero Energy Growth" projection. The seeming paradox is resolved by observing the difference between iea's and others' observation of trends. When GNP, population, productivity of the labor force, and other parameters are assumed to continue to grow at historically decreasing rates, and if energy prices escalate at about 2 percent per year in real terms (i.e., above inflation in the general economy), then the future looks very different. Earlier trend extrapolations had not taken such trends into account. Certainly more complex modeling efforts than iea's have been made. Still, iea's work highlights the importance that exogenous assumptions play in scenario analysis. As one might imagine, the implications for policy, such as whether and how soon to build a nuclear breeder technology, are drastically affected by the choice of such assumptions. Even when analysts (such as those at iea) heavily stress the use of electricity for replacing petroleum, a moratorium on nuclear power looks much less ominous for energy supply under the assumptions which produce a low energy demand projection.

1.3.8. Alternative Energy Demand Futures to 2010 (CONAES-1979)[12]

The National Academy of Sciences' Committee on Nuclear and Alternative Energy Strategies (CONAES), through its Demand/Conservation Panel, produced six projections of future (2010) energy use in the U.S. The Committee's original and central concern was to clarify the issue of urgency for breeder reactors. The purpose of the Demand/Conservation Panel (one of four panels established by the Committee) was to explore a wide range of plausible energy demand paths. Whereas other studies varied their scenarios by changing the variables of GNP, population growth, etc., the main difference among the CONAES' projections was the varied assumptions about energy prices.*

Assumptions regarding price in the six CONAES scenarios varied in the following way. One assumed that the price of energy over the next thirty years would decline at an annual rate of two percent a year. Another assumed that energy prices would not increase in real terms. Both we can dismiss because since the CONAES analysis was performed, price increases have occurred that would equal an annual rate of increase of two percent per year over the next thirty years if no other real price increases occurred (that is, if the price of energy simply increased at the rate of inflation in the general economy). Another scenario, "CONAES B," assumed a two percent annual real energy price increase. A variant on this scenario, "CONAES B '," uses the two percent price increase, but in the context of a more rapidly growing economy. CONAES A and A* scenarios are based on a four percent annual real price increase, the former without and the latter with strong nonmarket conservation incentives and regulations.

The results are striking (see Figure 1.1 and Table 2.1). The "A* Scenario" resulted in a demand of only 58 quads for the year 2010. Fifty-eight quads may be an unlikely future, but it is one that is plausible and without the need for serious sacrifice. The "A Scenario" which places the demand for the year 2010 at about the same as the 1980 demand, begins to look even more realistic. The "CONAES B Scenario" resulted in a year

---

Except two variations: one for nonmarket regulation of consumption (A), and another for higher GNP growth (B ').

2010 demand level of 94 quads, well below that of the EPP "Zero Growth Scenario" projection of 100 quads.

1.4. A Perspective on Future Energy Demand

Do the differences in opinion on our energy future constitute differences in principle? Thomas Jefferson once said that ". . .every difference in opinion is not a difference in principle." Given the distance between the high and low energy scenarios in the foregoing, can a consensus ever be reached? Does one or the other pole of opinion represent the pursuit "of completely abstract problems. . .transformed in some unforeseen manner in the artificial brain?"

To answer these questions, and to judge the divergent scenarios of future energy demand, we must turn to a detailed analysis of the ingredients of energy demand.

Les grandes personnes aiment les chiffres.

—Antoine de Saint Exupéry
Le Petit Prince (1943)

Chapter Two

The Elements of Energy Demand

2.1. Number Crunching

The author of *The Little Prince* made a serious accusation when he said that *grown-ups love numbers.* Saint Exupéry complained that when children tell grown-ups they have made a new friend, the grown-ups never ask any questions about essential matters. "They never say to you, 'What does his voice sound like? What games does he love best? Does he love butterflies?' Instead they demand: 'How old is he? How many brothers does he have? How much does he weigh? How much does his father make?'" With profound apologies to Saint Exupéry and his *petit prince,* we plunge into a whole mess of numbers.

The calculation of the coefficients for energy demand modeling is unceremoniously referred to as "number crunching." Yet it is the heart and soul of demand analysis and crucial to the judgment of energy demand studies. This chapter analyzes each of the important parameters and refers

Table 2.1. Summary Table: The Elements of Energy Demand Assumptions of the Major Demand Scenarios (Increase, in Percent Per Year, to the Year 2000)

Scenario	Population	Labor force	GNP	Energy productivity	Energy price (real)
1. FPC–1970	1.3	NG	4.0	NG	0.8[a]
2. Limits to Growth–1972	1.4	NG	3.4	NG	0
3. AEC–1973	NG	NG		NG	<1.0
4. EPP–1974	.75	1.0	2.73	1.25	0.7–2.8[b]
5. USBM–1975	.93	NG	3.25	.6	0
6. FEA–1976	.75	NG	4.58	1.3	0
7. iea–1978	.56	.7	2.7		2.0
8. CONAES–1979					
"A*"	.8	.8	1.8	NA	4.0[c]
"A"	.8	.8	1.8	NA	4.0[c]
"B"	.8	.8	1.8	1.5	2.0[c]

Source: Reference 7.

Abbreviations used: NG = not given, NA = not available.

[a] To the year 1990.   [b] Crude oil price only.   [c] To the year 2010.

back to the demand studies described in the prior chapter in order to evaluate them.

There are several main ingredients to energy demand. These include, to name a few, income, size of the labor force, rate of household formation, worker productivity, energy productivity, energy price, income, and cross elasticities. Projecting energy demand can be like trying to make a pie for which the recipe keeps changing. A change in the quantity of one key ingredient, like population, can require changes in all the others and can in effect alter the size of the pie. For reference the values assumed for the ingredients of demand for the major energy demand studies discussed in the following section are listed in Table 2.1.

2.2. Population

Population is one of the most important ingredients in energy demand growth. Population determines to a large extent the size of variables such as labor force, GNP, household formation, the number of automobiles, and others. While recent trends have all pointed toward increasingly lower fertility rates, population is very volatile, as the post-World War II baby boom made quite clear.

The fertility rate is the average number of births per female. The rate in 1975 was *1.8* children per woman in contrast to almost *four* children per

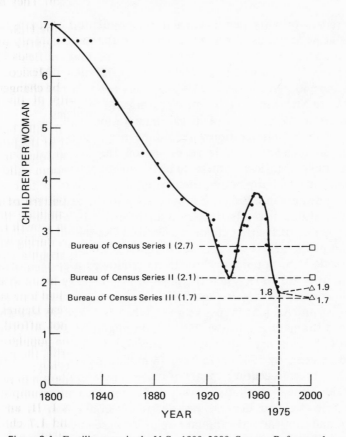

Figure 2.1. Fertility rates in the U.S., 1800–2000. Source: Reference 1.

woman in the 1950s. A graph of the fertility rate shows it to have declined steadily from seven children per woman in 1800. It dropped off sharply during the Great Depression and declined rapidly during the 1960s, marking the end of the baby boom (see Figure 2.1). Whether it will continue to decline or reverse itself is a matter of conjecture*; a consensus among demographers seems to be that the fertility rate will decline to 1.7 children per woman by the year 2000 because a substantial number of pregnancies are unwanted. It is significant to note that if illegitimate births to teenage girls were cut in half, the fertility rate would drop to 1.5 children per woman.[1]

It might seem paradoxical that while the fertility rate is less than the replacement rate of two children per woman, the population is still growing.

---

*The fertility rate rose ominously in 1979.

There are two primary factors causing the population of the U.S. to continue to grow at about .8 percent per year at the present. They are illegal immigration and the age distribution of the population.

Between 500,000 and 1,000,000 illegal aliens enter the U.S. each year and perhaps 60 percent remain.[1,2] The overwhelming majority of the immigrants are Mexican. The recent discoveries of large oil fields in Mexico could alleviate the problem of illegal immigration from Mexico. Whether this rate of illegal immigration will continue or perhaps be changed to legal immigration depends to a large extent on the health of the Mexican economy and on Mexican–American birth control policies.

Following the baby boom, the percent of women of child-bearing age increased not only in number, but in relation to the rest of the population. Therefore, although the fertility rate diminished, the absolute number of women capable of bearing children increased. As the boom children grow older, this phenomenon will decrease in importance.

Historic U.S. population growth has followed a pattern of generally declining rates. In 1800, when the population was 5.3 million, the annual rate of increase was three percent. The rate of population growth fluctuated between 2.8 and 3.1 percent per year until the Civil War, during which time it dropped dramatically to two percent. Immigration steadily swelled the ranks of U.S. citizenry. Although the growth rate never again went higher than 2.6 percent, by 1900 there were 76 million people living in the U.S., and the seventeenth century's fears of underpopulation had long since been reversed by Thomas Malthus' postulate. During the Great Depression, the fertility rate declined because people simply could not afford to have children, and thus the total annual rate of increase in the population fell to .7 percent. Population growth increased, of course, during the post-World War II baby boom but has since continued its historic decline.

The U.S. Bureau of the Census has projected population in the U.S. to three different levels, each corresponding to a different assumption about the fertility rate. These estimates are known as Series I, II, and III and assume that the fertility rate will average 2.7, 2.1, and 1.7 children per woman, respectively. Extrapolation of these rates yields a total U.S. population of 287 to 245 million in the year 2000.

Table 2.2 presents the resulting overall population growth rates of the three Bureau of the Census projections (1.2 to .56 percent per year) with rates assumed in the major demand studies. Unless the recent upturn in the fertility rate represents a new trend, the Census I and II projections overestimate the future U.S. population. Therefore, energy demand studies which use growth rates higher than those of the Census III projection probably overestimate. Even the Census III projection, with an assumed fertility rate of 1.7 babies per woman, could be too high. Note that all but the iea-1978 Low Scenario assumed the higher fertility rate.

Table 2.2. A Comparison of Population Growth Rate
Assumptions (Growth in Percent Per Year)

Scenario	Year Projected	
	2000	2010
1. FPC–1970	1.3[a]	—
2. Limits to Growth–1972	1.4	—
3. AEC–1973	NG[b]	NG
4. EPP–1974	.75	—
5. USBM–1975 (all scenarios)	.93	—
6. FEA–1976 and 1977	.75	—
7. iea–1978 (low scenario)	.56	.2
8. CONAES–1979 (all scenarios)	.8	.6
For reference		
Census Series I	1.2	1.15
Census Series II	.8	.6
Census Series III	.56	.2

Source: Reference 7.
[a] To 1990.
[b] NG = not given.

Several other surprises tumble out of this jumble of numbers and thus make it worth a close look. The Energy Policy Project (EPP-1974), which shocked the field with its low energy projection, used a population growth rate equal to that used by the FEA-1976 forecasters whose results were far higher. In this case both assumed a lower population growth rate than did the USBM-1975 and the FPC-1970, however. More surprising is that the CONAES study, which produced some very low (relatively speaking, of course) energy futures, assumed a higher population level than iea-1978, EPP-1974, and FEA-1976. Different assumptions about other parameters such as energy price and GNP growth, therefore, had more of an effect on energy projections than did population.

2.3. Labor Force

Determination of the labor force at future dates is important to energy modeling because it is a prime determinant of GNP. Growth in GNP is equal to growth in labor force times growth in labor productivity. The labor force is determined by the population size, the number of persons old enough to work (over 16 years of age in the U.S.), the number of persons seeking employment (the participation rate), the unemployment rate, and length of the work week. As the surge of humanity exploding out of the baby boom reached working age, the demand for jobs skyrocketed along with the demand for goods and services. That surge was a partial cause of the high rate of unemployment which recently went higher than seven percent, far above the average post-war rate of between four and five percent. The fertility rate, however, has dropped drastically over the last two decades, and thus this factor in the labor force will be lessened within the coming decade. The annual rate of increase in the labor force, which was 1.2 percent in the 1950s, 2.5 percent in the 1960s and early 1970s, should begin to diminish in the early 1980s. Series II and III population forecasts would show the labor force growth increasing annually at 1.9 percent until about 1990. Then the Series show it declining to .8 or .7 percent, depending on the fertility rate, and further slowing to .6 and .45 percent in II and III respectively. If the actual labor force turns out to be larger, however, then more workers turning out more goods which use energy both in their production and operation will drive energy demand higher.[1]

Further, young people are more likely than older persons to need a new car or a new house and appliances. With the 14–24 age group declining as a percentage of the total population from 15 to 10 percent by the year 2000, and declining in absolute numbers by more than six million, a major strain on energy supply will be relaxed.[2]

The greater participation of women in the work force will increase its numbers somewhat, although much of the increase in female participation may have occurred already. About 45 percent of American women are employed outside the home, compared with more than 78 percent of all men. One projection[2] envisions that the percentage of all women working will grow to 52 percent by the year 2000, compared with that of men at a slightly reduced level of 76 percent.

Mandatory retirement has caused a drastic decline in the percentage of men working past the age of 65. In 1950, 43 percent of all men ages 65 and over still worked, compared with only 21 percent today. By now this issue has become highly politicized, and federal mandatory retirement ages have been dropped altogether. The impact of this policy change on the labor force is difficult to predict, although it almost certainly will reverse to some extent the declining participation of older workers. Additionally, the percentage of the population 65 and older will probably increase from about 10 to 12 percent.[2]

Labor force and unemployment assumptions of the major demand studies are given in Table 2.3. Note that the FPC-1970, the Limits to Growth-1972, the AEC-1973, the USBM-1975, and the FEA-1976 studies were not sufficiently detailed to take into consideration the size of the labor force. These studies thus failed to take into consideration a major relaxation of stress on energy supplies, the relaxation that will come with a diminished rate of growth in the labor force.

Although the labor participation rate may override and obscure the significance of unemployment, unemployment as a factor of energy

Table 2.3. A Comparison of Labor Force Growth Assumptions
(Growth in Percent Per Year)

| | Year Projected | | |
Scenario	1985	2000	2010
1. FPC–1970	NG[a]	NG	—
2. Limits to Growth–1972	NG	NG	NG
3. AEC–1973	NG	NG	NG
4. EEP–1974			
Technical fix	1.4	1.0	—
Zero energy growth	1.4	1.0	—
5. USBM–1975	NG	NG	NG
6. FEA–1976 and 1977	NG	NG	—
7. iea–1978 (low)	1.9	.7	.45
8. CONAES–1979 (all scenarios)	1.9	.8	.6

Source: Reference 7.

[a] NG = not given.

demand may take on added importance now that the U.S. has in the *Humphrey–Hawkins Act* a national policy calling for reduction of unemployment rate to four percent. It is not certain that lower unemployment would increase industrial output and, consequently, energy demand, because it might simply result in shorter working hours and more leisure for the employed rather than an increase in total working hours. Most of the demand studies we evaluated, including the iea-1978 and the CONAES-1979, assumed unemployment levels of about five percent.

2.4. Labor Productivity

The other half of the GNP equation, labor productivity, depends on many things: how well trained the work force is, the modernity of capital equipment and its replacement rate, the quality and availability of freight transportation, communications, and financial services, government policies, and the availability of raw materials and energy.[1]

Since 1940, productivity has increased at an average annual rate of about 1.6 percent. Although this rate was 2.2 percent per year from 1950 to 1965, the last decade has seen it decline to .9 percent due to several factors including a slower turnover in capital equipment (due in part to difficulties in raising investment capital, and perhaps to a poor management and worker attitude).[3]

If any of the major energy demand studies we have analyzed, other than iea-1978 and CONAES-1979, made assumptions regarding labor productivity, they have done a good job of keeping them secret.

2.5. Energy Productivity

The issue of energy productivity, the quantity of energy required to produce a dollar's worth of GNP, is central to projecting energy demand. Until recently, it was widely assumed that the two were related in a fixed ratio. A graph of energy per dollar of GNP of several countries for a given year would seem to verify such an assumption because the relationship between national wealth and national energy consumption is linear. The poorest countries huddle in the low energy/low income corner, whereas the wealthy nations stretch in the opposite direction in a line which seemingly shows one-to-one energy–GNP relationship. That this relationship has been misinterpreted is one subject of Chapter Three.

Turning the ratio of energy to GNP on its head, to obtain its reciprocal, also provides a convenient measure of energy productivity. The GNP to energy ratio observed as a trend over the last 50 years shows a surprising result: energy productivity in the U.S. has actually been increasing despite a steady (until 1973) decline in energy prices. Increased technical efficiencies have developed despite the shrinking incentive to substitute ingenuity in the form of energy-saving capital for energy. Energy productivity has also resulted partly from the fact that the U.S. has been moving toward a service economy wherein sectors such as education, travel, and health care grow faster than manufacturing.

In terms of constant dollars per BTU consumed, energy productivity has increased at a rate of about .8 percent per year since 1955 while real energy prices declined at an annual rate of about 1.8 percent. Depending on the assumption made for the price elasticity of energy, which we discuss below, increasing fuel prices would logically seem to spur even faster increases in energy productivity. The assumptions made in this regard by the major modeling efforts are presented in Table 2.4.

Note that many of the studies ignored potential improvements in energy productivity altogether. Naturally, these had the higher demand projections, with the exception of the USBM-1975, which did assume some (small) energy productivity growth.

2.6. Gross National Product

The concept of GNP is supposed to serve as a measure for the output of the economic system, and it has come to be a surrogate for the well being or quality of life of nations. To reiterate, GNP equals the number of workers working times the average labor productivity. Such a measure inevitably includes things like the sale of Saturday Night Specials to would-be robbers and the medical costs of pollution. GNP figures assume that if

Table 2.4. A Comparison of Assumptions for Growth in Energy
Productivity (Rate of Increase in Percent Per Year)

Scenario	Year projected		
	1985	2000	2010
1. FPC–1970	NG[a]	NG	—
2. Limits to Growth–1972	NG	NG	NG
3. AEC–1973	NG	NG	NG
4. EPP–1974	NG	NG	NG
5. USBM–1975	1.25	1.25	—
6. FEA–1976 and 1977			
Business as usual	.6[b]	—	—
Conservation	.6[b]	—	—
7. iea–1978 (low)	2.1	1.3	1.0
8. CONAES–1979[c]			
"Scenario B"	1.5	1.5	1.5

Source: Reference 7.

[a] NG = not given.

[b] To 1990.

[c] An *outcome,* not an assumption.

you do your laundry and I do mine, nothing happens to GNP, but if I pay
you to do mine and you pay me to do yours, then GNP goes up.

Ward and Dubos said in 1972 that [4]

> Ten years from now, our concept of GNP may include little boys swimming in
> the Delaware or Volga, days gained for industry that were formerly lost to
> bronchitis and head colds, the drapes and clothes no longer due for the cleaners,
> the number of days without smog in central cities, the vigor and serenity which
> comes from walking to work in friendly neighborhoods, the lessening in police and
> prison costs, the rise in leisure as people begin to like sitting beside their city
> sidewalks, to find their neighborhood parks alive with rock or Bach or the police
> brass band, to walk in countrysides and wildernesses that are cared for and
> protected. (pp. 141–142)

Bad indices, like bad theories, are not overthrown upon being proven
weak or even false; they are overthrown when something better comes along
to replace them. The key question for our energy future, then, is "Can GNP
(or the quality of life) increase without a simultaneous and equivalent increase in energy use?" As we have suggested, there is considerable evidence
that it can.

A comparison of the GNP assumptions, which is one variable all of the
major studies deal with, is offered in Table 2.5.

The iea–1978 and CONAES–1979 figures reflect lower assumed growth
rates for labor productivity and labor force. The higher assumptions of four
percent or more are curious. GNP growth has averaged only 3.4 percent per
year since 1945, and only 2.1 percent for the decade of the 1970s. And those
were times of higher population growth.

Table 2.5. A Comparison of Assumptions for GNP Growth (Rate of Growth in Percent Per Year)

| | Year projected | | |
Scenario	1985	2000	2010
1. FPC–1970	4[a]	—	—
2. Limits to Growth–1972	3.4	3.4	
3. AEC–1973			
4. EPP–1974 (all scenarios)	4.06	2.73	—
5. USBM–1975	3.4	3.25	
6. FEA–1976 and 1977 (all scenarios)	4.58	—	—
7. iea–1978 (low)	3.6	2.7	2.5
8. CONAES–1979			
"B"	2.3	1.8	1.6
"B ′ "	3.5	2.6	2.1

Source: Reference 7.

[a] To 1990.

2.7. Other Factors

Once an energy modeler has a population, he or she must put it into houses and cars and provide not only enough commercial space for them to buy food, furniture, Christmas presents, and books, but also all the other amenities. It would be easier, from the modeler's perspective, to lump everyone into high rise apartments and be done with it. But Hannah Arendt wisely warns us to pursue human, as opposed to abstract, objectives.

There were about 74 million residential units* in the U.S. in 1976, with about 2.9 persons living in each one. One trend seems to be toward more people living alone. The iea-1978 study, in consideration of this trend, forecast that while there would be only 15 percent more people in the U.S. by the year 2000, there would be 42 percent more households. In comparison, the CONAES-1979 study called for a .7 percent per year higher rate of increase in the number of households than in population growth. Other trends include more space, better insulation, more bathrooms, more efficient appliances, and more multifamily dwellings.

An important concept in modeling energy consumption is market saturation, the point at which consumers have all of a product they are willing to buy. It is a point in which case the market is limited essentially to replacement purchases and to any expansion of population. It appears that

---

*A residential unit equals one living unit, be it an apartment, detached single family dwelling, or whatever.

such a point is being approached for automobiles, refrigerators, cooking ranges, water heaters, air conditioners, and other basic appliances.[5] Televisions had reached an average of 1.4 sets per home by 1970. We are now rapidly approaching a situation in which we have one car per licensed driver, and furthermore, people now spend about 45 minutes in a car every day. Over the past 30 years, then, what has occurred has been a saturation in what might be termed the basic comforts of life for a majority of Americans. The new consumer items like computer games or stereos for the home or car tend to be much less energy intensive. Thus as items like these begin to move up their saturation curve and become a relatively larger component of economic activity, the energy/GNP ratio would tend to decrease.

Commercial space is a sort of surrogate for growth in the service industry. Most of the major studies neglect this aspect of the energy future, with only iea-1978, EPP-1974, and CONAES-1979 venturing estimates. Projections hover around a two percent per year rate of increase. The FPC-1970 report did mention commercial growth, but in connection with the opportunity to expand electrical output by providing lighting for the parking areas in new shopping centers associated with urban sprawl.

Price elasticities measure the near-term as well as the long-run consumer responses to changes in energy price. Elasticities are important modeling tools in the short run, that is, perhaps to a time as distant as 1990. Beyond that, the concept is less useful than estimates of efficiencies which can be derived with economically applied engineering improvements. (For a comparison of assumptions regarding energy price increases, see Table 2.1.)

On the subject of price elasticity, the FPC-1970 authors (Reference 3 of Chapter 1) asserted that a

> . . .reversal in the trend of the (declining) price paid for electricity must be taken into account, along with the probable price performance of competitive sources of energy. . .while customer acceptance of many electrical uses show little price elasticity, certain large and growing markets (such as space heating) are price competitive. . .*it would be shortsighted to view electric energy growth in terms of cost–price adjustments,* or to question the rate of growth without considering the nation's overall energy needs and supplies. (pp. I-3 to I-5; emphasis added.)

Limits to Growth ignored price elasticities and did not consider improvements in engineering efficiencies. USBM-1975, like the FPC report, was candid: "1974 prices (were assumed). The forecast does not incorporate price effects. The available data are insufficient for such consideration." (p. 25)

EPP did incorporate price effects in its modeling effort. The Hudson–Jorgenson model which EPP-1974 employed indicated that a 40 cents per million BTU price increase would, over the long term, lead to

- o a 7.8 percent decrease in energy consumption
- o .6 percent *more* employment
- o .5 percent more capital consumption*
- o 1.0 percent greater inflation
- o .5 less GNP

Energy prices current at the time of the preparation of EPP's *A Time to Choose* included a cost of $8.50 per barrel of fuel oil or about $1.50 per million BTU. Thus, a 34 percent price increase would result in a 7.8 percent decrease in demand, an overall price elasticity of about − .23.

Iea's main analysis did not rely on econometric modeling, but instead took the engineering approach to determine future efficiency improvements. As a check on its work, though, iea performed a separate macroeconometric modeling effort in which "GNP was related to labor, capital, technological change, and energy."[1] The total of these estimates yielded an overall price elasticity of − .41 in the Low (Demand) Case, and − .29 in the High Case.

CONAES-1979 divided its modeling effort into the short run and the long run. Econometric modeling was performed out to 1990 because it was felt that existing data on price elasticities were plausible for that period. Beyond that time, CONAES-1979 analysts believed that the more defensible approach was economic–engineering modeling. The (long-term) elasticities CONAES-1979 assumed for the period to 1990 follow:

Residential	−.7
Nonresidential Buildings	
o electricity	−.62
o gas	−.75
o oil	−.63

The percentages by which the energy intensity of important end uses can be reduced by the year 2010 (as assumed by CONAES-1979) are shown in Table 2.6 for the scenario of energy price increasing at a rate of two percent per year ("B Scenario").

The technical possibilities behind these estimates and the potential for going beyond them form a major portion of Part III. It should be pointed out that some of the aforementioned savings do include operational changes, but that none of them require any but minor behavioral changes.

---

*This was the assumption for the relationship of energy and capital in the first Hudson–Jorgenson Model. Dale Jorgenson now apparently believes that capital and energy are complements, not substitutes.

Table 2.6. CONAES-1979: End Use Energy Intensities
in the Year 2010

End use	Percent reduction in energy intensity (new products in 2010 compared with 1975)
Residential buildings	37
Commercial buildings	40
Government/educational buildings	55
Agriculture	15
Aluminum production	37
Cement manufacture	37
Chemicals production	22
Construction	35
Glass manufacture	24
Iron and steel production	24
Paper production	29
Other industry	25
Automobile transport	(27 mpg)[a]
Vans; light trucks	(21 mpg)[a]
Air passenger transport	55
Truck freight transport	20
Air freight transport	40
Rail transport	3

Source: Reference 8.

[a] National fleet averages of newly manufactured vehicles.

2.8. Some Judgments

2.8.1. Criteria

Two key criteria can be applied to evaluate energy demand analyses: (1) reproducibility, and (2) validity of assumptions. Reproducibility as a criterion demands that a study's results be obtainable by other analysts working with the same data, assumptions, and tools.[6] A trend extrapolation cannot be reproduced if none of the major trends are identified and, like any experiment, cannot be considered conclusive. The selection of one assumption over another is to make a prediction, and, as Alvin Weinberg, director of the Institute for Energy Analysis (Oak Ridge, Tennessee) and former director of Oak Ridge National Laboratory, has warned, "Never make a prediction until you are very old. . .you might live to see it not come true."

2.8.2. Reproducibility

Perhaps we have placed a gun on the mantle which will not be used. To use the criterion of reproducibility seriously would eliminate from consideration half of the aforementioned studies. FPC-1970 states that its analysts performed two separate econometric analyses, but besides the macroeconomic assumptions previously cited, none of their methodology was explained; neither was it referenced so that others could obtain it. The AEC-1973 study did not divulge its methodology. Similarly, the USBM-1975 described neither the mathematical formulation behind the forecast nor its sources of information. The FEA-1976 PIES model is described in detail, but not one assumption is referenced. The Limits to Growth-1972, EPP-1974, iea-1978, and CONAES-1979 studies each described methodology and assumptions in detail.

2.8.3. Validity of Assumptions

Rapid change in social and political situations makes evaluation of assumptions difficult, a problem aggravated by variation in the mathematical formulation of models. Modelers may tie price increases only to supply or only to demand, or to both. The results would differ in each instance. Tying price to demand, for example, would lead to lower demand. Similarly, certain parameters such as GNP can be either assumptions or results. That is, they can be assumed exogenously and function as independent variables and "drive" the model, or they can be endogeneous and operate as dependent variables. It is not always easy to tell which is the case in a model. More importantly, energy demand scenarios can call for quite different policy measures depending on the assumptions and constraints cranked into them. Forecasts are often no more than mirror images of their premises.[7] When it comes to energy demand modeling, the saying "Garbage in; garbage out" is a better than average aphorism.

The key parameters that drive these models and determine their futures are energy price, GNP, and population. The FPC-1970 estimated a real electricity price increase of about 18 percent between 1970 and 1990. Already, this estimate is too low by about 80 percent, and prices are not likely to stabilize soon. The Limits to Growth-1972 and USBM-1975 envisioned no real increase (actually a real decline) in overall energy price to 2000. The AEC-1973 forecast a small increase, while FEA-1976 assumed two cases: one, that the price of oil would not increase in real terms, and two, that it would decrease 40 percent. As Gary Wills, (author of *Inventing America: Jefferson's Declaration of Independence*) once said, "There is no accounting for what some people think."

GNP, if it is the product of labor force and productivity, will not grow as fast in the aggregate as it has in the past, although *per capita* GNP growth could be faster than ever. It is our judgment that GNP growth forecasts at four percent or more per year are far too high. FPC-1970, *Limits to Growth,* AEC-1973, EPP-1974, USBM-1975, and FEA-1976-7 have each assumed such a rate. Thus, these studies are seriously and incorrectly biased toward high energy demand projections. The iea estimate of 3.53 percent per year until 1985 seems high. The CONAES-1979 assumption of 2.7 percent per year is probably the best estimate.

Of the three Census Bureau projections, Series II was the most often chosen. Only one study, the FPC-1970, opted for Series I, the highest rate of growth. In our opinion, Series I is not really credible, and even Series II seems cautiously pessimistic. The FPC-1970, *Limits to Growth,* AEC-1973, EPP-1974, USBM-1975, FEA-1976-7, and CONAES-1979 studies each assumed population growth rates above or near that of Series II. Only iea-1978 chose Series III, which many demographers believe to be the most likely. Poor assumptions for population growth, therefore, have driven energy demand estimates higher than might realistically be expected in all but the iea-1978 projection. CONAES-1979 scenarios A*, A, and B, projecting energy demand lower than iea-1978, would have forecast still lower demand in 2010 with less "conservative" population assumptions (i.e., for a slower rate of population growth).

2.9. Conclusions

The issue of saturation of energy demand for certain goods and services arises, as we noted in the last chapter. The FPC-1970 forecasters let shine through their enthusiasm for increasing electricity consumption in a section on outdoor lighting:

> Improved outdoor lighting is generally recognized as being both desirable and important in improving the quality of life, but the Nation has only begun to consider large-scale outdoor lighting as an amenity of urban living. Studies have shown that increased illumination levels reduce automobile accidents, that increased lighting in downtown commercial areas attracts more customers, and that properly lighted streets discourage crime. In 1967, there were 3.7 million miles of streets and roads in the United States, of which about .5 million miles were in urban areas. Even in those areas most of the lighting was by low-power obsolete incandescent lamps. In addition to residential street lighting and outdoor parking area lighting, *the potential outdoor lighting market also includes the 3.2 million miles of streets and roads in rural areas,* some of which are lighted only at major intersections and village centers. (pp. I-3 to I-11; emphasis added.)

Perhaps it is just crankiness, but to us it seems of some value to leave areas where the night sky is dark enough to see stars.

Forecasts which place 1985 demand higher than 85 quads per year and year 2000 demand higher than 100 quads appear to be anachronistic. The FPC-1970, Limits to Growth-1972, AEC-1973, EPP-1974, USBM-1975, and FEA-1976 almost surely overestimate future energy demand.

Why have we concerned ourselves with bad forecasts? One might as well bully the weatherman. Scenarios such as FPC-1970, the USBM-1975, and especially the FEA-1976-7 are still taken seriously, though, and too many high level energy policymakers still look mainly at the future of supply, and not of demand. In their personal models of the future they connect the level of, say, gas production with higher gas prices, but do not give sufficient attention to the next logical step which is to inquire what happens to demand in response to price. High demand projections have fixed the concept of energy inelasticity firmly in the minds of our leaders. Consequently, 90 percent or more of energy research, development, and demonstration (RD & D) goes into mammoth supply schemes. As theory cautions, the models' results may call for policies contrary to what is needed.

Thus, pursuit of a high energy future begins to take on the aura of a pursuit of something abstract, an endeavor divorced from real human needs, an obsession with *les chiffres*. As a result, we may squander billions of dollars on uncertain, desperate energy supply schemes to produce quantities of energy that could have been conserved at a fraction of the cost. In the process, we may well lay waste much of our natural environment, and needlessly destroy many lives. The time lost on such misadventures may hinder us from effecting a workable energy strategy and thus threaten our very survival. Such a scenario, as opposed to a logical, stepwise approach to the future, would seem to constitute a serious difference in principle.

> . . .it is obviously contrary to the law of nature, however it may be
> defined, for a child to command an old man, for an imbecile to lead a
> wise man, and for a handful of people to wallow in luxury while the
> starving multitude lacks the necessities of life.
>
> —JEAN JACQUES ROUSSEAU
> *Discourse on the Inequality Among Men* (1754)

> *The only thing that could really affect the [energy] situation [of the poor*
> *countries in the world] would be for the industrial countries. . .to limit*
> *their oil consumption deliberately for the benefit of the developing world.*
>
> —MANS LONNROTH ET AL.
> *Energy in Transition* (1977)

Chapter Three

The U.S. in World Society

3.1. The U.S. and the Third World

There is something frightening in the statistic that one-third of the
world's people are dependent upon firewood for fuel. In the eastern
highlands of Africa, individuals frequently spend more than a half-day
journeying to obtain the firewood that once grew outside their doors. In
other cold mountainous regions, in the Andes, the Himalayas, and many
other places around the world, forests recede in ever-widening circles
around villages. Denuded hills erode and silt the rivers and add to the
danger of landslides and flooding. Because of extensive wood cutting in
Colombia, partly for firewood, the Anchicaya reservoir has filled with silt
in less than seven years, and the multimillion dollar hydroelectric plant it
was built to support now can run at only one-third of its intended capacity.
As firewood in Nepal becomes less available, the cow dung that farmers
have traditionally returned to agricultural fields now must be burned as

cooking fuel. Not only is the resulting air pollution damaging in the extreme to human health, but severe food shortages are exacerbated.[1,2,3,4]

Villagers find themselves spending more and more of their income for fuel. While the price of kerosene has escalated at a disastrous rate, in many places the price of firewood has tripled or quadrupled since the early 1970s. *Thus even renewable fuels must be conserved,* must be husbanded in their production, and used efficiently in stoves, not in fireplaces, tin cans, or open pits. Note the following description of conditions in 1978 in Bangladesh[5]:

> An unsealed road, inches deep in dust in the dry season, a sticky bog in the wet, leads along a levee from the town of Sherpur to Shapmari. Across a ribbon of pounded dirt separating the rice fields is the first house in the village. Its owner is the local barber, who practices his trade in the dry season in whatever shade he can find under the scattered trees. His charges vary from half a taka to one taka, or approximately three to six cents. He makes on the average day twelve cents.
>
> At harvest time he can earn the same amount, and two meals a day, by helping in the fields. Apart from the marginal supplement that his elder children may be able to earn tending the animals of the richer farmers, or collecting cow dung and leaves, this is the basic income on which he supports his wife and nine children. His stove is a basin-shaped hole dug in the earth. For fuel he uses leaves and twigs. He has, of course, no lavatory or running water. (p. 64)

An obvious conclusion is that business is not good for barbers in Bangladesh. But business is not good for millions in Bangladesh. About 60 of the 84 million people in that country, the most crowded in the world, earn less than 25 cents per day. As much as a third of the world's population may be suffering similar desperation if not despair. Consider a similar picture from India[6]:

> . . .an experience I had in south India in July of 1965. I was walking along a dusty road just outside of Mysore City one afternoon when I came to the outskirts of a very small satellite village. In front of one of the huts on the edge of the village I noticed the ever-present pile of round cakes, made from a mixture of cow-dung and straw, baking in the hot afternoon sun—the typical fuel of that area. Across the field in front of the hut an Indian woman was walking towards her home, carrying on her head two large jars of water from the village well some distance away. Even my limited knowledge of the lifestyle of these people told me that before long this woman would be using some pounded grain to make the chapattis, the flat bread cakes, to be baked over the cow-dung fire and served with a bit of rice or some vegetable as the major, if not the only, meal of the family for the day. (p. 4)

Another statistic, often quoted, imposes itself on all such discussions: the U.S. consumes 30 percent of the world's oil and yet has only 6 percent of the world's population. Is the assumption fair, often implicit in the use of this statistic, that the U.S. (and the developed world) gorges itself to the profound detriment of half of mankind? This accusation has been struck home with the ungloved fist of Thomas Pynchon's fiction[7]:

> Taking and not giving back, demanding that "productivity" and "earnings"
> keep on increasing with time, the system removing from the rest of the world these
> vast quantities of energy to keep its own tiny desperate fraction showing a profit:
> and not only most of humanity—most of the world, animal, vegetable, and
> mineral, is laid waste in the process. The system may or may not understand that
> it's only buying time. And that time is an artificial resource to begin with, of no
> value to anyone or anything but the system, which sooner or later must crash to its
> death, when its addiction to energy has become more than the rest of the world can
> supply, dragging with it innocent souls all along the chain of life. (pp. 480–481)

Pynchon was nominally writing about a chemist named Kekule, I. G. Farben, and the discovery of the benzene ring—organic chemistry in the Third Reich, but he clearly meant the analogy to be general. The point has been made more broadly and more politely in scientific journals[8]:

> In the worldwide search for routes to a juster and more sustainable society it
> has become clear to many observers that a peace in which the world is divided ever
> more rigorously into haves and have-nots is neither just nor likely to be sustainable,
> whether the basis for division is social, economic, or. . .seemingly technological.
> Such division not only defeats itself in the long run; even worse, it is wrong.
> (p. 57)

Is this fair? Does not the tiny fraction using the large amount of energy feed the world and supply a cornucopia of manufactured goods including power plants, trucks, copper wire, stoves, steel girders, and aluminum ingots? The answer, as we shall see, is no; the world does not depend on America's huge energy appetite for its food or for its goods.

3.2. The U.S. and the Developed World

Darmstadter[9] exploded the above myth (along with some others as they relate to energy conservation in the U.S. economy). Addressing the issue of whether the U.S. uses much of its enormous energy consumption to supply the world with energy-intensive manufactured goods, he offers the surprising fact that the U.S. is actually a net importer of energy in terms of the energy embodied in goods. That is, while the U.S. does export energy bound up in the manufacturing process of the goods it exports, it actually imports slightly more of such embodied energy in addition to the direct import of fuel. As for the question "Doesn't the U.S. feed the world (and therefore require copious quantities of energy)?" Darmstadter cites the fact that agriculture* in the U.S. consumes only 1.5 percent of the national

---

*Energy in agriculture is taken here as the amount of purchased energy used on the farm.

energy budget, and that relative to paper, steel, and chemicals manufacture, agriculture is not an energy-intensive industry.

Several questions then arise. Does the U.S. differ greatly from the industrialized world? Is our economy very different, or is it our lifestyle, or just a history of low energy prices (and therefore of low incentive to conserve)? What can we learn by comparing ourselves with other developed nations, and how can we use what we learn in molding our energy future?

A chart of GNP versus energy use in the world's nations would show a relatively consistent trend. The higher the energy consumption goes, the higher the GNP. Ratios of energy to GNP (E/GNP) are not good indices to compare energy efficiency, unfortunately, because of international differences in climate, geography, economic structure, lifestyle, and import–export balances. International comparisons of energy use are useful only if they are made on a relatively disaggregated level as, for example, on the level of automobile miles traveled per capita and distance between major cities.

3.2.1. The Swedish Example

Much has been made of the comparison of Sweden's energy use to our own. Too much, some would say. The interest in comparing Sweden to the U.S. is generated by its nearly equal (actually higher) standard of living, but far lower rate of energy use. Some interesting contrasts and a few surprises are listed in Table 3.1.

The figures in Table 3.1 are surprising. While Swedes use their automobiles 60 percent as much as Americans, their cars use only 60 percent as much fuel per passenger mile. Thus, the average Swede uses only 36 percent as much gasoline or diesel in a car as the average American. Because it is faster to get from Malmo to Stockholm by train than by car, and because the trains are pleasant and run on time, and because American mass transit when available is not always so pleasant, and for other reasons, mass transit energy consumption for the average Swede is 235 percent that of an American. While that is no surprise, it is surprising to learn that although Swedes have a colder climate, and keep their houses warmer than we do, their residential energy use per capita is 20 percent lower (per person, per degree day, per square foot) than ours. Industrial energy use is very different. Sweden produces four times more paper per capita than the U.S. because of a relatively larger forest resource and good export markets. Still, the efficiency of paper production is greater than ours by more than 10 percent, because the price of energy has been higher. Price has been kept high by taxes because Sweden must import almost all its fuel, and the high price stimulates efficient use.[9]

Table 3.1. Sweden/U.S. Contrasts in Energy Use:
Swedish Use Per Capita by End Use Category, as a Percent of U.S. Use

Category	A[a] Usage factor	B Efficiency factor	C Total
Automobiles	60	60	36
Mass transit	290	80	235
Urban trucks	95	30	28
Residential space heating	170	50	81
Appliances	NA[b]	NA[b]	55
Commercial space conditioning	130	60	78
Heavy industry		60–90	92
Paper	420		
Steel	110		
Oil	50		
Cement	135		
Aluminum	50		
Chemicals	60		
Light industry	67	60	50
Electrical generation (thermal)	30	75	23

Source: Reference 9.

[a] Column A, usage factor (e.g., relative percent of passenger miles per capita) times Column B, efficiency factor (e.g., relative fuel efficiency in automobiles) = Column C, total per capita consumption (as a percent of per capita U.S. consumption).

[b] NA = not available.

3.2.2. The U.S. vs. Nine Developed Countries

An analysis[10] of energy use among nine developed countries reached five major conclusions:

1. Transportation differences account for nearly one-half of the variation in the energy/GNP ratio among countries, with the remainder divided up among residential/commercial energy industries, and energy transportation. Differences in total energy use in manufacturing were small because greater efficiencies in Europe are offset somewhat by greater output of raw materials compared to the U.S.

2. U.S. households spend a larger fraction of their incomes on direct energy purchases than do European. . ... Not surprisingly, U.S. prices were the lowest.

3. After adjusting for climate, the American citizen uses 40 percent more energy for space comfort than the average European. Part of the reason is the larger, detached homes in the U.S. *vis-à-vis* the continent. Moreover, full house heating is only saturated in North America and Scandinavia. Efficiencies were significantly better in Sweden, with Canada's lying between Sweden's and the U.S.'s.

Table 3.2. International Comparisons: Energy Use

Factors	U.S.	Canada	France	West Germany	Italy	The Netherlands	United Kingdom	Sweden	Japan
Energy prices (lowest prices = 1)	1	2	4	5	9	6	8	3	7
Passenger miles per unit Gross Domestic Product (GDP) (most miles = 1)	1	5	9	3	2	6	4	6	8
Percentage of passenger miles accounted for by cars (highest = 2)	1	2	8	4	7	5	6	3	9
Energy consumption per car–passenger mile (least efficient = 1)	2	1	4	8	5	7	9	6	3
Cold climate (coldest = 1)	7	2	5	4	9	3	5	1	8
Size of house and percentage single family (largest = 1)	1	1	6	6	8	5	4	3	8
Extractive industry GDP as percent of total GDP (highest = 1)	2	1	8	4	6	9	3	5	7
Industrial GDP as percent of total GDP (highest = 1)	7	8	2	1	6	4	5	9	3
Ratio of industrial energy consumption to industrial GDP (highest = 1)	2	1	9	8	6	5	4	3	7
Degree of energy self-sufficiency (highest = 1)	2	1	7	5	8	3	4	5	9
For reference:									
Energy/GDP ratio (highest = 1)	2	1	9	6	7	3	4	5	3
Energy per capita (highest = 1)	2	1	7	5	9	4	6	3	8
GDP per capita (highest = 1)	1	3	4	5	9	6	8	2	7

Source: Reference 9.

4. In the transportation area, total travel, efficiency of each mode, and (to a lesser extent) actual mix of modes vary widely. The auto dominates everywhere (except in Japan), but higher urban densities, higher auto costs, and higher fuel costs and other measures have kept the absolute level and relative share of heavily subsidized public transit at two to six times the U.S. level per capita. Higher U.S. freight consumption results mainly from the fact that the U.S. moves larger volumes of freight farther but also because the U.S. system appears to rely overall more on rail. The higher consumption, however, is due to structural or historical or geographical or political reasons and it appears to have little to do with energy.

5. In industry, it takes more energy in the U.S. to create a given product than in most countries in Europe. Obviously other factors, such as the cost of capital and labor, influence the choice of technologies used in industries.

Table 3.2 provides some interesting international comparisons of energy use in industrialized nations. Nine countries are ranked ordinally for each of 13 parameters affecting energy use, including energy price. The lowest ranking (i.e., 1) along each factor would indicate heavier energy use, while a rank of 9 would indicate the least heavy. For example, the nation with the lowest energy prices (the U.S.) would have rank 1, and so far as this variable is concerned, would be the most inclined toward energy use. That nation with rank 9 in the "cold climate" row (Italy) would have the mildest climate, and the least energy demand because of this factor.

Note that among this group of nations, the U.S. is second only to Canada in energy self-sufficiency. In Gross Domestic Product (GDP) per capita, the U.S. is first. In energy use per capita, Canada exceeds the U.S. Italy ranks ninth, while Japan, even with its growing economy, ranks eighth, although in GDP per capita it is seventh.

To omit a conclusion from Table 3.2 would do violence to our purpose. International comparisons of energy use show that there is much technical flexibility and conservation potential within present U.S. energy use patterns, and to other countries, provided that economic incentives and time are allowed to play a role.[10]

3.3. The Equality Gap

Perhaps Pynchon's vision of apocalypse will not come to pass. Still, there is the question of what is to be done about the huge material disparity of more than 10 to one between the desperate and the wealthy. How to bring the two groups closer together equitably and peacefully, quickly and wisely, may be the greatest question not only of our time but of all time. How to allocate oil, one of the most basic resources necessary for the physical well-being of the six billion people who, barring famine, war, or a miracle, will coexist on this planet in the year 2000, is central to this question.

3.3.1. The Oil Gap

The desperation that emanates from an oil shortage is a troubling thing. Oil companies send people into the harshest of environments to find crude oil: to the Arctic where at times one could lose a finger by taking off a mitten and simply touching a piece of metal; to the North Sea, where 60 foot waves are not uncommon and where a $16 million drilling operation may find nothing but dry rock. We have been willing to foul the oceans with oil in the process of shipping it. Nations have been willing to go to war, as England did (or, rather, tried and failed to do) at Suez in 1956, to risk their prestige and power to protect their supply of oil. Egalitarian governments such as our own, functioning under "enlightened self-interest" have supported the worst of tyrants to get oil.

There is another sort of gap: the gap between the developed countries' appetites for oil and their supplies. The developing countries possess 82 percent of the world's proven oil resources. Under what terms they will allow the developed countries to use their oil and how much they will reserve for their own use in development is an intriguing question.

There are two points to be made in this connection. The first is that it is not necessarily sensible for the developing countries to follow the same development path as did the developed countries. This is true for several reasons, but in our present context it is pertinent to point out that much industrial development has been predicated on the existence of cheap oil. There is not enough cheap oil available for the third world to follow the same path. As Lonnroth *et al.* put it, "There simply is not enough cheap oil going around. The poor countries must abstain from a line of development which cries every bit as much for cheap oil as ours has done."[11]

The second point is that for the developed countries, different forms of energy are more or less equivalent (from a utilization standpoint). Industrialized nations have sufficiently developed infrastructures so as to be able to function with any type of energy, given time for adjustment, of course. Developing nations generally do not enjoy such flexibility. However, a less developed infrastructure is required for the use of oil than, for example, nuclear fuel or coal. Oil is easy to transport, store, and burn. It may therefore be the energy form of choice for the most rapidly developing countries. (Fortunately, as we discuss in Part II, synthetic liquids from biomass may be substituted for oil.)

It is the developed countries, however, which currently are gorging themselves as if there were no tomorrow. We are therefore faced with the situation Lonnroth *et al.* picture for us:

> The industrial countries are rapidly consuming the oil that exists in the developing countries. At the same time there are large energy resources in the industrial countries. It may be questioned whether energy in the form of oil (liquid fuel) is not necessary to enable the developing countries to industrialize as we have done in the industrial countries. It is therefore plausible to argue that all faith in the industrialization of the developing countries, in the sense familiar to us today, will be buried at the same rate as the industrial countries consume this finite resource. (p. 83)

3.3.2. How Much Oil? The CIA

What is the outlook for oil production? We are getting a bit ahead of ourselves, because this question is one we want to explore thoroughly in Part II. But in relation to the U.S.'s role in a world desperate for oil, a glance ahead is in order.

A short paper, the *International Energy Situation: Outlook to 1985,*[12] garnered a good deal of attention after its release in 1977 with its bad news forecast. The unnamed authors, employees of the U.S. Central Intelligence Agency (CIA), predicted a serious world oil shortage by 1985, a shortage of no less than seven million barrels of oil per day (see Figure 3.1). The world currently produces about 65 million barrels per day. Preceding this gloomy prospect was to be a price break, at which time the impending shortage would touch off sharp price increases.

The one bright aspect of this study is that it could be seriously wrong. Its demand projection was a rather crude macroeconomic extrapolation. Economic growth in the U.S. was assumed to be 4.5 percent per year until 1985 and 4.0 percent thereafter. Western European countries were assumed to grow at a rate of 3.5 to 3.8 percent annually, while developing countries' economies grew at 6.5 percent. Annual energy growth for the world would average 2.9 to 3.4 percent, and in the U.S., 2.8 percent. Oil consumption in the U.S. would climb from 40 quads in 1979 to 49–56 quads per year. If the CIA were right, the 1985 U.S. oil demand level would be as great as the year 2010 projection in the "CONAES C" scenario. But whoever, if anyone, is right, Lonnroth's "We have an unstable situation" will be the understatement of the decade.

3.3.3. Oil Supply and Demand: A World View

The *Workshop on Alternative Energy Strategies (WAES)*[13] took a more sophisticated approach to modeling world energy supply and demand. The analysts, collected from many different parts of the world, based their

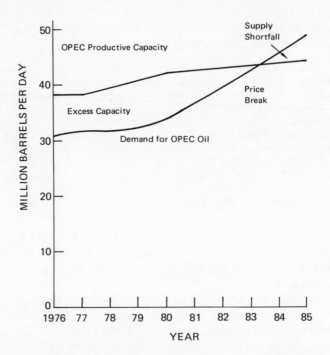

Figure 3.1. OPEC oil: The supply/demand gap. Note: A more realistic plot of this graph would deflect the demand curve downward and the supply curve upward. Conservation should help reduce demand beyond what the CIA assumed. Similarly, the CIA assumption for the contribution of Mexican oil was probably too low. The net result would be a much smaller "short fall." Source: Reference 12.

analysis to the year 2000 on the following assumptions:

o   Three different price levels:
    •   No increase from 1976 levels through 2000
    •   Rising price, from $11.50 per barrel of oil, to $17.50 by 1985, then no increase until 2000
    •   Falling price, to $7.66 by 1985 (this scenario was taken only until 1985)
o   Two levels of world economic growth:
    •   3.5 percent annually to 1985, then 3 percent to 2000
    •   6 percent to 1985, then 5 percent per year to 2000 [North America's rate in these scenarios would be either 4.3 or 3.0, to 1985; 3.7 or 2.6 to 2000. Non-OPEC countries have annual economic growth rates of 6.2 or 4.2 percent per year until 1985, and then slow to 4.6 or 3.7 percent annually between 1985 and 2000. This contrasts with the goal of the United

Nations for an average annual growth rate in the Less Developed Countries (LDCs) of 6 percent.]

Note that the price assumptions were chosen rather arbitrarily, either to rise or fall 50 percent above or below 1976 oil price levels. More importantly, note that they are already hopelessly out of date. While econometric modeling designed to find the supply–demand equilibrium and price solutions would have been useful, the scenario approach is illuminating. WAES concluded that:

> . . .the supply of oil may fail to meet increasing demand well before the year 2000, most probably between the years 1985–1995, even if energy prices rise 50 percent above current levels. . .failure to recognize the importance and validity of these findings and to take appropriate and timely action could create major political and social difficulties that could cause energy to become a focus for confrontation and conflict. . . . (p. 5)

The WAES study thus offers yet another bad portent. A bright spot is, again, that they perhaps overestimated energy demand growth in North America, as well as perhaps for the Western European countries, especially in light of oil price increases in the 1970s. "This does not," as Lee Schipper has said, "spoil the analysis."[14] He was implying that the crisis would simply be delayed in time a bit (no small advantage) and, therefore, is not to be disbelieved. In this regard, we might recall that Cassandra's ability to predict the future was accompanied by one small, irritating curse. No one would ever believe her.

3.3.4. Oil and Energy Demand: A Western View

A more widely known if slightly less detailed study which corroborates the WAES finding is the *World Energy Outlook* (not to be confused with the Exxon study of the same name) published by the Organization for Economic Cooperation and Development (OECD). This study assumed that economic growth in the OECD countries would range between 4.2 and 4.3 percent annually through 1985. Resulting European petroleum import requirements came to about 77 quads, or 35 million barrels per day, almost double the current U.S. total petroleum demand. The scenario assumed successful implementation of all conservation policies in effect in 1976.[15] Lower GNP growth rates, estimated to average between 3.6 and 3.8 percent per year, would reduce import needs by nine percent. Recall that U.S. economic growth as assumed by CONAES was only 2.7 percent per year. While the U.S. is certainly not Europe, and vice versa, the economies of the two are linked in important ways, indicating that one of these estimates for economic growth might be wrong.

3.3.5. Oil and Energy Demand: An Oil Company View

Exxon Corporation's *World Energy Outlook* (1978)[16] makes several very interesting (if incredible) assumptions:

o   A zero real oil price increase through 1990
o   Oil continues to supply incremental demand for energy throughout the world
o   OPEC nations will meet increasing demand by increasing production

Exxon makes other assumptions, namely that conservation will be supported by the governments of the developed nations, and that world economic growth will continue, but at slower than historic rates. None is so tenuous, though, as those for price and of the goodwill and cooperation of OPEC.

Exxon foresees the economy of the U.S. growing at 3.4 percent per year in 1990, with that of the combined developed nations at 3.7. While the economies of the U.S. and the industrial countries grew at rates of 3.6 and 4.7, respectively, from 1965 to 1973, many things have changed. Recall from Chapter Two that productivity and the labor force are bound to grow more slowly than during the 1960s. European productivity has been high because its capital equipment, the steel furnaces, the factories, etc., are newer and more efficient than those in the U.S. Rapid European capitalization since World War II has led to greatly increasing productivity. As this capital equipment ages, productivity increases will be harder to win. Since GNP growth is the product of growth in labor force and productivity, the Exxon (and other) GNP estimates could thus be too high. Because Exxon assumes that GNP and energy use go hand in hand, its projection of 330 quads of world energy demand (compared to 243 quads in 1972) could be seriously high.*

3.3.6. Energy Demand and the World Economy: An Academic View

Wassily Leontieff, the inventor of the "input–output" method of economic analysis, spearheaded a projection of world economic activity including energy use. The study, sponsored by the United Nations and published in a book modestly entitled *The Future of the World Economy,*[17] dealt with, among other things, the income gap between the rich and poor countries of the world. Using different growth assumptions, Leontieff addressed the equity questions we have been flailing at in this chapter.

---

*Only energy sold on commercial markets in non-Communist countries. Exxon's 1979 *Outlook* assumed sharply reduced growth rates for developed countries.

Table 3.3. Leontief's Estimate of Future Annual Growth Rate
in Energy Demand (1970–2000)
Industrial vs. Developing Countries

Area	Oil	Gas	Coal
Industrialized countries	4.1	3.0	5.0
Developing countries	9.1	10.4	6.4

Source: Reference 17.

Two scenarios were developed. In one the present rich-to-poor income gap of 12 to one narrows to seven to one by the year 2000, and in the other the gap stays the same. The wider chasm would result if present trends are extrapolated, that is, if the *status quo* prevails. This scenario foresees developed countries getting richer at an annual rate of 4.5 percent (but with a *per capita* increase of 3.5 percent). This rate is based on population growth in the developed countries of one percent. The narrower chasm could result if population in the developed world grew only .6 percent per year, and if the population increase in the Third World were only two percent, not the 2.5 as projected by historical extrapolation. These changes would allow Third World economic development to increase at a 6.9 percent rate rather than six percent, according to Leontieff *et al*. The implications for growth in energy use under this latter scenario are shown in Table 3.3.

In the context of the analyses presented in Chapters One and Three, the energy demand growth rates for the developed countries seem astronomically high. Even Exxon predicted only a rate of 2.8 percent for energy growth in the developed nations. Also, it seems that Leontieff implicitly assumed a zero real price increase for oil or perhaps even a falling price.

3.3.7. The World Energy Dilemma

There are (at least) two implications here. Leontieff and others have probably overestimated energy consumption in the developed world; if so, the Third World could close the income gap even further than Leontieff predicted. On the other hand, increasing energy prices could crush the developing countries. Whatever the result, it appears imperative that the conservation of energy in the developed world be vigorously pursued for the well being of poor nations. It is in our own long-term self-interest to conserve oil in order to give the poor a chance. To a very large extent, we can provide this chance without economic sacrifice. With volition, we could provide a lot more.

3.4. The Implications for U.S. Energy Policy

The implications of the analysis we have presented in Part I are at once conflicting, and therefore both disconcerting and reassuring. While it is reassuring that the U.S. will probably need less energy than many originally believed, it is troublesome that developing countries are going to need so much. It is reassuring that energy conservation can reduce energy demand without requiring a sacrifice of the amenities we enjoy; yet, it is troubling that we are expending so little effort on energy conservation. And it is reassuring that conservation can extend our energy supplies and thereby provide us with the time required to make an orderly transition to sustainable fuels in a relatively low energy future; yet, troubling in the extreme that until we can make that transition we will be dependent on the goodwill and stability of unstable nations, and that we will be competing for their favors against a significant part of the world that is growing and sometimes desperate.

The implications for U.S. energy policy are that we must move quickly, but logically, to make this transition. Unhappily, we know too little about how to begin that transition insofar as energy supply is concerned. It will take the decade of the 1980s to organize ourselves to start manufacturing acceptable alternate energy systems. Happily, we do not have to panic, since we can make enormous strides toward reducing energy demand by increasing energy productivity. This potential is the subject of Part III.

In Part II we turn to an exploration of energy supply possibilities. Energy availability and price determine how much conservation is economical, where conservation is most useful, and how much time we actually have to get the job done.

My candle burns at both ends;
It will not last the night;
But ah, my foes, and oh, my friends—
It gives a lovely light.

—EDNA ST. VINCENT MILLAY
Poems (1923)

Part Two

Energy Sourcery
Energy Supply, Price, and External Costs

Introduction to Part Two

An enlightening projection of future energy demand and supply comes from John Yegge,* who places energy in the perspective of geologic time. Starting with the appearance of the first animals on Earth, an event which occurred roughly one billion years ago, Yegge translated time into a single calendar year with January 1 marking the appearance of those primeval forms, and midnight, December 31, representing the present. Not much happened until July, when oil was formed. This phenomenon was, of course, marked by the proliferation of small

*Personal communication with Chandler, 1974. Currently with Pacific Science Center, Seattle, Washington.

63

aquatic invertebrates. The hydrocarbon chains which composed these animals became oil. Great forests appeared in September; they became coal. The dinosaurs appeared in early October, but were gone by mid-November. In between, the Appalachians were formed. The first mammals came on the fifth of December. Man finally appeared about 2:20 a.m. on December 31! The last ice age ended about 11:52 p.m. America was discovered at 11:59:45 p.m., 15 seconds before midnight. The industrial revolution began at 11:59:53, and Drake struck oil in Pennsylvania at 11:59:56. Midnight is the present. Oil is to be gone at 12:00:02, and coal is to be exhausted at 12:00:05. Thus, the Earth's fossil resources are gone in the twinkling of an eye.

But what does Yegge mean by "gone"? That there are no more economically recoverable reserves, or that there is absolutely no more to be had at any price? And what does economically recoverable mean? At today's prices? Furthermore, what did he use as the basis of his resource estimates?

Yegge's purposes were, of course, merely pedagogical. He often pointed out that he was not concerned about the accuracy of the resource estimates he used. Double the estimated resources, he argued, and the implications for society were ultimately the same. Some day we must make a transition from finite fossil fuels to sustainable energy resources.

The ultimate fallacy that still pervades our thinking is that exponential growth will continue indefinitely. Under that supposition any act of conservation (for example, doubling of energy productivity) only buys two or three decades. It is imperative that we shift our thinking to terms of a steady state for energy demand, with economic gains to be had in other ways.

A small revolution has come to supply as it has to demand, however. Just as higher prices make conservation more attractive, they make more expensive supply options feasible. Thus, we have both the Department of Energy and Congress drawing plans for plants to make synthetic gas from coal for $6.00 to $10.00 or more per million BTU, when not long ago gas cost only $0.25 or less per million BTU. Recently, a popular magazine[1]* published an article which suggested that copious quantities of natural gas can be had for about $6.00 per million BTU or less from geopressured zones, deep-lying deposits, tight sands, coal mines, and shales. The author of that article went so far as to suggest that a methane hydrate layer under Siberia's tundra could supply the world's gas needs for 3000 years.

*This reference can be found in the list of references for Chapter Four.

Examination of the complexities of energy supply are crucial to an assessment of energy demand for several reasons. In a special sense, supply always equals demand. More importantly, the *price* of energy (which is intimately related to supply) determines the level of conservation which is economically feasible and to a large extent the level of final demand. Also, the total amount of energy used determines the amount of external costs incurred. Very important is the possibility that, as energy supplies shift over time, the demand sector can also shift, especially if given sufficient lead time. In the long term, there is a large substitutability in the buildings sector of electricity or solar energy for petroleum fuels. Alternative sources such as alcohol, methane, hydrogen, and electricity all are usable in automobiles if given fifteen years or so for transition.

We do not intend one of those turgid accounts of every conceivable energy source. We are interested in the all important liquid and gaseous fuels, oil and natural gas, and their potential substitutes, solids, synfuels, nuclear and solar energy. And we are vitally interested in the associated environmental and health costs of the use of these energy sources. But, primarily, our intent is to provide a context of energy supply and price against which the demand scenarios in Part I and the conservation options in Part III can be evaluated.

Chapter Four explores the availability of oil and natural gas, which presently supply 75 percent of U.S. energy demand, and the outlook for price.

Chapter Five examines the potential of solid fuels, with primary emphasis on coal and uranium. Synthetic liquids and gases manufactured from coal, oil shale, or biomass and solar energy options that may be substituted for oil and natural gas are also examined. How the cost of these supplies will compare with oil is a major topic.

Chapter Six is an in-depth look at the other costs associated with energy use, the environmental and human health factors.

Chapter Seven is a synthesis of Chapters Four through Six and an examination of the implications these chapters bear for U.S. energy policy.

In 1885 the U.S. Geological Survey indicated there was little or no chance of finding oil in California and in 1891 this was expanded to include Kansas and Texas. The maximum future supply of U.S. oil was envisioned to be 22.5 billion barrels in 1908, and in 1928 the United States had oil supplies for only 13 years at then-current consumption rates. By 1949 the end of the U.S. oil supply was almost in sight.

—HERMAN KAHN ET AL.
The Next 200 Years: A Scenario for America and the World (1976)

Chapter Four

Liquids and Gases, The Crux of the Matter

4.1. Oil

The U.S. currently uses about six billion barrels of oil per year (or 15 million barrels per day), but it produces only 3 billion barrels. Before World War II, the U.S. supplied approximately 70 percent of the world's oil, but now supplies 15 percent while using 30 percent. If the U.S. relied solely on its proven reserves of about 40 billion barrels, domestic oil would be gone at current consumption rates in about seven years.

Proven world reserves total about 650 billion barrels,* compared with

---

*Note that this estimate includes only about 26 billion barrels of proven *reserves* for Mexico. Mexican *resources* are far higher, but it is by no means certain that these can be recovered.

annual world production of about 20 billion barrels. Estimates of the world oil resource range between 1.5 and 3 trillion barrels. (Recall that a barrel equals 42 gallons.) Before we proceed, however, it would be helpful to explain the complex terminology of oil.

Proven reserves: Already discovered and accurately known oil recoverable at current prices with present technology.

Probable reserves: Projected discoveries in existing fields, recoverable at current prices with present technology.

Resources: Concentrations of oil including discovered and undiscovered quantities which are now or may someday be economically extractable.

The complex methodology for forecasting oil supplies deserves some elucidation, too, for out of these estimates comes the information crucial to making plans for an energy future. Almost all government energy policy including oil production subsidies, conservation policy, breeder reactor development, environmental control, price regulation, national security, and all energy research and development is affected by oil supply projections. Thus much is at stake in the accuracy of these projections.

4.1.1. Crude Oil Estimates for the U.S.

The task of estimating the crude oil resource is made somewhat easier in the U.S. because so many drill bits have penetrated potentially oil-bearing sedimentary rock, and because such extensive geophysical prospecting has been done. Reserve and resource estimates are basically statistical. They relate the rate of discovery of reserves to the cumulative footage of exploratory drilling (the dR/dE ratio). M. King Hubbert first used this technique in 1956.[2] His and other, frequently larger estimates are displayed in Table 4.1. A graphic display of his analysis is reproduced in Figure 4.1.

Today a consensus seems to center around the estimate of 250 billion barrels, and Hubbert's estimate in 1956 seems "almost prophetic."[3] Gregg Marland[4] believes Hubbert's original estimate of 150–200 billion barrels of oil is closer to being correct than the current "consensus" figure of 250.

Hubbert's technique was to extrapolate a declining dR/dE. His famous bell-shaped curve of U.S. oil production which (in 1956, heretically) showed production peaking in 1973 and declining thereafter was drawn by integrating the annual increments of discovery over time. The area under the curve represented 170 billion barrels of oil.

Table 4.1. Estimates of the Ultimately Recoverable U.S. Oil Resource[a]

Date of estimate	Total recoverable (billion barrels)	Estimated by
1956	150–200	M. King Hubbert, U.S. Geological Survey
1961	407–507	Senate Committee on Interior & Insular Affairs (Lasky)
1962	885–1000	McKelvey (U.S. Geological Survey), letter to National Academy of Sciences of July 20, 1962
1963	650	Duncan and McKelvey (U.S. Geological Survey)
1965	400	Hendricks (U.S. Geological Survey)
1969	168	Hubbert, "Energy Resources," in *Resources and Man,* Chapter 8, National Academy of Sciences
1971	432	National Petroleum Council–American Association of Petroleum Geologists (Ira H. Cram)
1972	420–2250	Theobald, Schweinfurth, and Duncan, U.S. Geological Survey Circular 650
1974	215	Hubbert, "U.S. Energy Resources; A Review as of 1972," U.S. Committee on Interior and Insular Affairs, U.S. Senate, 1974, Committee Print Serial No. 43-40 (92-75) Part I, US6P0.
1975	236	National Academy of Sciences, National Research Council, Committee on Mineral Resources and the Environment, 1975, Washington, D.C.
1975	230	J. D. Moody and R. E. Geiger, "Petroleum Resources: How Much Oil and Where," *Technology Review,* March–April 1975, pp. 39–45.
1976	252	Exxon Company, USA, *U.S. Oil and Gas Potential,* 1976, Pubic Release, Houston, Texas.
1976	187	Gregg Marland, *A Random Drilling Model for Placing Limits on Ultimately Recoverable Crude Oil in the Coterminous U.S.,* Published in *Materials and Society,* Pergamon Press, 1978, Great Britain.

[a] Total equals oil already removed plus that which is ultimately recoverable.

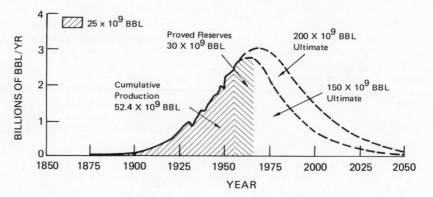

Figure 4.1. Peak crude oil production: Hubbert's prediction for the U.S. Source: Reference 2. (BBL = barrels.)

In contrast, the Zapp Hypothesis (used by Cram in 1971; see Table 4.1) forecasts the discovery of far more oil. Zapp, in effect, extrapolated the dR/dE from the early 1960s without a further decline, arguing that increased knowledge of oil geology would increase the level of the resource found.[5] His conclusion was clearly wrong, however, because the dR/dE has declined steadily since his forecast (see Figure 4.2). Barry Commoner, in the *Poverty of Power,* implied that Cram used a similar method to arrive at his 432 billion barrel estimate.* Marland[3] suggests that authors may be "picking" their resource estimates to fit the needs of their advocacy.

Marland's contribution is far more substantial than cynical, however. His *Random Drilling Model*[4] offers a means of evaluating these conflicting predictions. His model assumes simply (1) that oil is discovered in proportion to the amount of oil which remains to be found and the area which remains to be searched, and (2) that this "random rate" can be related to a time-dependent variable which describes our knowledge of where oil is found and the area of the U.S. under which oil may be discovered. The model borrows data which indicate that the distribution of oilfields by size is lognormal. In other words, the number of separate oilfields which would have to be discovered and pumped increases exponentially in relation to the amount of oil one produces. Such data are not surprising, for we might intuitively believe that there can be only a few East Texas-size (giant) oilfields and very many small fields. To increase the oil resource, we will have to resort to drilling an increasing number of wells which produce progressively less oil.

*Cram actually derived his estimate by summing separate (optimistic) regional estimates. Cram's estimate assumed 60 percent recovery.

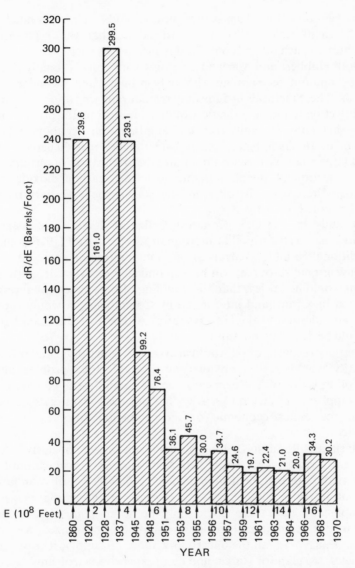

Figure 4.2. The discovery rate of oil (dR/dE) as a function of time, 1860-1970. Source: Reference 4.

It is difficult to dispute Marland's conclusion that if, in the U.S., there is a great deal more conventionally recoverable oil than 200–250 billion barrels, a tremendous quantity must exist in very small fields. To refute Marland's conclusion, we must deny that oilfields are lognormally distributed, and/or believe that the oil industry drills only for very small fields at random. If we dismiss these notions and assume that much oil does exist in small fields, then we must admit that the exploration and production of such oil will considerably increase its cost.

Both Hubbert and Marland deal only with conventionally recoverable oil, that amount recoverable with only primary and secondary recovery methods. These methods include only regular pumping and water or natural gas reinjection to increase the flow of oil by increasing the pressure in the oil strata. Such conventionally recoverable oil amounts to about one-third of all the oil originally in place. Recall that Cram assumed 60 percent recovery. Such a large recovery rate would require the application of tertiary recovery by the injection of steam, miscible chemicals, carbon dioxide, or other methods of removing the oil from the porous rock. Tertiary recovery, it should be noted, is costly.

A study by the U.S. Congress Office of Technology Assessment[14] estimates that even with oil at more than $30 per barrel (1976 dollars) only an additional 50 billion barrels of oil, and addition of about 20 percent to the conventional resource, will be economically recoverable. Although this amount would be far less than the doubling of the conventional resource as estimated by Cram (and interpreted by Commoner), it would represent a significant addition to the U.S. resource. A total of 1.5 million barrels per day could be derived from this resource by 1985.

Tertiary recovery of oil would also exact its environmental price. Steam generation would result in noxious emissions, already a problem in tapping heavy oil in California. Chemical injections could cause contamination of water supplies. Tertiary oil recovery has not been investigated thoroughly enough for either economic or environmental risks to be excitedly promoted.

Marland's Random Drilling Model is not totally conclusive. After all, it deals with statistics reporting petroleum exploration which until the last few years did not include drilling beneath 5,000 feet. Drills can now reach depths of nearly 25,000 feet. Also, ever since we learned that Hegel proved philosophically that there could only be seven planets and that Lord Rutherford stated that nuclear fission could never be utilized, we have tried to be cautious. With all the uncertainty about oil supply, it seems prudent, nonetheless, to plan for scarcer and more expensive petroleum.

4.1.2. Oil Production Estimates for the World

The U.S. has become increasingly dependent on world oil supplies and production since 1960. Crude oil production of the major exporting countries of the world in 1977 is shown in Table 4.2.

It should be obvious by now that future world oil production may depend more on politics than geology. The oil producing countries listed in

Table 4.2. Annual World Oil Production (1978), excluding the USSR

Country	Million barrels per day	Quadrillion BTU	Percent
Algeria	1.2	2.5	2.0
Iraq[a]	2.6	5.5	4.4
Kuwait	2.1	4.5	3.6
Libya	2.0	4.2	3.3
Qatar	0.5	1.0	0.8
Saudi Arabia	8.3	17.6	14.0
United Arab Emirates	1.8	3.8	3.04
Subtotal: Arab OPEC	18.6	39.1	31.0
Ecuador	0.2	0.4	.3
Gabon	0.2	0.4	.3
Indonesia	1.6	3.3	2.6
Iran[a]	5.2	11.0	8.7
Nigeria	1.9	4.0	3.2
Venezuela	2.2	4.7	3.7
Subtotal: Non-Arab OPEC	11.3	24.0	19.0
Total OPEC	29.9	63.3	50.0
Canada	1.3	2.8	2.2
North Sea	1.5	3.2	2.5
Mexico	1.2	2.5	2.0
Total OPEC, Canada, Mexico, North Sea	33.9	71.8	57.0
U.S.	8.7	18.4	15.0
Eastern block nations	17.0	36.0	28.5
Total World	59.6	126.0	100.0

Source: Reference 10.

[a] Output has been drastically reduced by war.

Table 4.2 (see also Figure 4.3) are politically unstable to some degree. Iran and Iraq remain in turmoil, and Saudi Arabia worries about an Iranian-style revolution. All the Mideast in fact is a powder keg. Nigeria recently held free elections, but memory of the cruel Biafran war and the genocide that followed it remains fresh. Indonesia is stable only for tyranny. Mexico suffers a terrible schism between the rich and poor and also a devastating rate of population growth. Canada and Venezuela alone of our suppliers seem stable. Though (and perhaps because) both countries exhibit democracies perhaps stronger than ours, they too may see benefits in limiting their output.

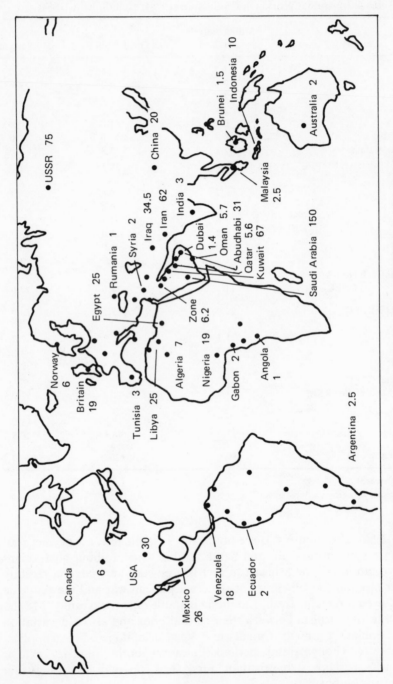

Figure 4.3. Publicly stated proven oil reserves, by country (in billions of barrels). Source: Reference 11. The figures shown for Mexico are subject to upward revisions. Unlabeled dots: ≤ one billion barrels.

To talk of politics is not to say that geology is unimportant. There is a production rate above which extracting oil from a field will damage it by reducing pressure too quickly. Extracting oil too quickly will leave much oil unable to percolate toward the well hole through the porous rock in which it lies. The maximum rate at which oil can be produced without such damage is called the maximum sustainable production rate. During 1977, many of the countries listed above had production rates averaging 20 percent below this maximum. This "shut in" production was a result of a temporary oversupply of oil in the world market, which, of course, disappeared with the Iranian revolution. Future oil availability, though, will depend on this rate (which is approximately 6.6 percent per year of the proven reserves) and on the addition of new reserves. Note that Mexico, which recently announced discovery of massive oil resources, produced less than all the major exporting countries except Qatar. Also note that the U.S. is not only the world's largest importer of oil, but one of its largest producers of oil. Its 1977 production of approximately three billion barrels was equal to its import of three billion barrels, and U.S. production was exceeded by only the USSR and Saudi Arabia.

Much has been made recently of the addition of 100–300 billion barrels of resources in Mexico. Mexico's discoveries are not proven reserves, however, and may be hard to retrieve. Perhaps only ten percent is recoverable at today's prices and technology. On the other hand, Mexico might become a new and even bigger Saudi Arabia. Other Mexicos are possible, but unlikely, and we might ask who is willing to gamble his or her nation's destiny on such odds? As one study[6] put it,

> Nations that continue to increase their oil consumption in the hope that more optimistic estimates will prove correct risk losing time to adjust their energy consumption patterns if their optimistic expectations are not met.

4.1.3. Strategic Petroleum Reserve

It is common wisdom that to reduce U.S. dependence on foreign oil is to increase our security and improve our economy; it is less well accepted that these same ends of security and economic health would suffer from a complete withdrawal from the world oil market on our part. To a large degree, however, global interdependence is a positive force, and imported oil is still cheaper than most of the alternatives. So, to panic and to embark upon a program of rapid energy independence could actually defeat our efforts to enhance our security and economic well being.

Still the threat of embargo gives rise to a rational fear. This being the case, Edward Krapels[7] investigated the extent to which an oil embargo would threaten us (see Table 4.3). His conclusions, particularly in regard to the value of a strategic petroleum reserve, are illuminating.

Table 4.3. Embargoes: Adequacy of Import Countries' Oil Stocks

A. Number of days to deplete reserves

Size of embargo (percent of imports lost)	Days required to deplete stocks (stocks = 60 days of imports)
7	no depletion
10	1200
15	720
20	352
25	240
30	180

B. Size of embargo vs. size of supply loss

Source of embargo	Size of embargo required to create a 20% supply loss	Size of embargo required to create a 30% supply loss
Saudi Arabia; Kuwait; United Arab Emirates; and Libya	73%	not possible
All Arab-OPEC	60%	90%
All OPEC	36%	54%

Notes and Assumptions

- Imports equal 60% of import countries' supplies.
- Import countries include: U.S., West Germany, France, Japan, Italy, and the Netherlands.
- Curtailment of 10 percent is imposed in each country in the event of an embargo (except, of course, for a 7% supply loss).
- In the U.S., stocks of commercial and strategic oil products equaled 50 days and 10 days, respectively, of oil imports.
- The 1973–74 Arab-OPEC embargo reduced western or importing countries' oil supplies by an average of 5%.

Source: Reference 7.

Krapels asked the question, "To what extent would oil suppliers have to embargo oil deliveries to Western nations in order to exhaust planned strategic petroleum reserves?" The planned reserves generally include an amount of oil equal to a 45 day supply. Table 4.3 details Krapels' answers. These include:

o   An embargo equal to that of 1973–4 (about a five percent total reduction) would have little effect for four years.

o   The Arab producing nations would have to reduce exports to the West (including Japan) by 50 percent for one whole year to exhaust the petroleum reserve.

o   All exporting nations would have to reduce production by 30 percent to eliminate the reserve within one year.

In the U.S. the reserve will consist basically of crude oil (as opposed to refined products—a reserve mix of crude and refined products might be more useful) stored in empty salt domes on the Gulf Coast. It is targeted to total one billion barrels. A problem in these scenarios should be evident, however. The reserve will not be fully in place before 1985. In 1981, it totaled only 200 million barrels, although commercial stocks of petroleum products equaled about 1.3 billion barrels, or 72 days of imports.

4.2. Natural Gas

Natural gas, the wonder fuel, burns blue-hot in the energy controversy. Because it is so clean and versatile, and because it has been so cheap, some of the most heated debates have occurred over its price and supply. Consumer advocates assert that its price should be regulated to reflect its cost of production plus a reasonable rate of return on investment for the gas industry, and they contend that higher prices will not increase the production rate.[8] The industry assures us that if we deregulate the price of gas we will be supplied with all the gas we want.[9] Whatever the merits of these arguments, Congress has voted to end the long regulation of gas prices in steps to 1985 when the price will be more or less decontrolled.* With this in mind we can ask, "What is the future of supply and price of natural gas in the U.S.?"

Untapped reservoirs of gas (the coastal geopressurized zones, the Western tight sands, deep-lying deposits such as the Andasco basin, Devonian shales, and Appalachian coal seams) plus the gas demand-diminishing effect of fuel switching and improved utilization efficiencies could cause yet another oscillation back to natural gas in the U.S. fuel consumption pattern. What are we to make of the claims of enormous quantities of natural gas?

For context, we should remember that natural gas currently furnishes the U.S. with about a quarter of its annual energy supply. That equaled about 20 quads out of 76 in 1977.[10] (The U.S. produces about 3.5 billion barrels of oil equivalent of gas each year.) Most of the official demand studies examined in Part I foresee the role of natural gas diminishing and not expanding. In the world context, natural gas is of far less importance, supplying almost no energy to Scandinavia or Japan. The Netherlands, drawing on the Groningen fields discovered in the early 1960s, obtains almost half of its energy from gas, however. Additionally, the gas associated with the North Sea oilfields will increase the importance of gas somewhat in Northern Europe.

*Gas from reservoirs that were being produced before 1977 are never to be deregulated, however.

Table 4.4. Estimates of the World Natural Gas Resource by
Region

Region	Billion barrels of oil equivalent	Quadrillion BTU
North America	280	1624
Western Europe	77	446
Middle East	270	1566
Other Noncommunist	348	2018
Communist	425	2465
Total World	1400	8120

Source: Reference 11.

4.2.1. U.S. and World Gas Resource Estimates

A chronicle of estimates of the U.S. gas resource would read with the same intrigue as that of oil. Rather than agonize across that history, we will simply report that a Hubbert analysis extrapolating a declining dR/dE resulted in an estimate of 1050 trillion cubic feet of gas as the ultimately recoverable U.S. resource. This total translates into a little more than 1050 quadrillion BTU, or 184 billion barrels of oil equivalent. Zapp later extrapolated the dR/dE rate to level off and decline no more, and thus estimated the existence of 2650 trillion cubic feet (2650 quads, or 457 billion barrels of oil equivalent). Zapp, again, has been proven wrong by further decline in the gas discovery rate.

A more recent estimate by Moody and Geiger[11] puts the total North American gas resource at about 1624 quads. Reserves and resources for the world (by region) are shown in Table 4.4 (see, also, reference 12).

Not all of North America's gas belongs to or will be available to the U.S. If Hubbert was right and the total U.S. resource is 1050 trillion cubic feet (1078 quads), and the U.S. has already produced almost 900 trillion cubic feet, then the current annual conventional natural gas production rate of about 20 trillion cubic feet will decline sharply over the next couple of decades. (The ratio of reserves to production is about 11.) The price of gas, of course, would escalate predictably.

Forecasting the natural gas resource* is even more difficult than for oil for several reasons, of which two stand out: (1) gas occurs both in

---

*Almost all of our past experience in gas exploration and development has been in association with oil. Thus, when we add the effect of exploration specifically for gas, the estimates of the gas resource could rise appreciably.

association with crude oil and separately (60 percent is unassociated gas)[6]; (2) alternative "unconventional" sources of gas could potentially hold an enormous supply which might be produced for less than or equal to the cost of synthetic gas from coal.

4.3. Unconventional Sources of Gas

4.3.1. Gas in the Coastal Geopressurized Zone

During the Cenozoic geologic era, which began about 60 million years ago, large quantities of sediment accumulated along what is now the Gulf Coast of the U.S. Overlapping deltas of silt, sands, and clays containing organic matter, probably associated with a marine environment, were buried along with large quantities of salt water. Increasingly, pressure built up on this trapped water and organic material, which, because of the high temperatures and impermeability of the clays, converted the organic solids to methane which became dissolved in the brine. The buildup of sediments continued causing growth faults which disjointed the strata and buried the sedimentary rock as deeply as 50,000 feet. As the water trapped within these fractured layers became saturated with methane gas, some of the methane escaped into caps in the overlying rock. It is from these caps that we obtain gas along the Gulf Coast today. Certain recent estimates of the resource that lies deep, saturated in brine aquifers, have ranged as high as 100,000 trillion cubic feet or 100,000 quads. Enough to supply the entire present U.S. energy demand for more than 1,000 years.[1,11,13]

If this resource does exist, the trick will be to get it out. Drilling to depths of 30,000 feet is possible, but an enormous quantity of water would have to be brought to the surface. One 42 gallon barrel of brine, for instance, might contain about 50,000 to 100,000 BTU in 50 to 100 cubic feet of gas. A barrel of crude oil, on the other hand, contains 5.8 million BTU. One estimate has placed the minimum economically feasible production rate per well at 40,000 barrels of brine per day, others at 100,000.[1,13] Presently, one test well is producing 10,000 barrels of water per day from an aquifer 12,600 feet deep. The temperature of the brine is 240°F, high enough for industrial process heating, and is pressurized at 11,000 pounds per square inch (atmospheric pressure at sea level is 14.7 pounds per square inch). This well, the Department of Energy's Edna Delcambre #1, produces 50 cubic feet of gas per barrel of water, or 50,000 BTU per barrel. Water at deeper levels may contain more gas, perhaps 100,000 BTU at 20,000 feet, or up to one million BTU at 30,000 feet. In any case, tremendous quantities of water must be extracted causing at least two problems: (1) surface subsidence, and (2) salt-water disposal. The U.S. Geological Survey[13] has calculated that if on-shore surface subsidence is limited to only three feet, then only five percent of the water can be removed. Subsidence would of course cause less of a problem offshore, and half the geopressurized aquifers do lie beneath the Gulf of Mexico. The disposal problem remains, though: the brine could not simply be dumped in the ocean because it contains not only heat but salt in concentrations ten times greater than sea water and possibly other contaminants as well. Perhaps a great deal of the gas can be extracted, possibly even cheaply enough and in large enough quantities to obviate for some decades the need for synthetic liquids and gases from coal. Still, however, uncertainty is large. The price of gas would undoubtedly be higher than at present. Even the most conservative estimates, on the other hand, project this resource to add about 132 quads (132 trillion cubic feet) to our gas supply.[9]

4.3.2. Gas from Devonian Shale

All along the Appalachians from New York to just south of the northeastern boundary of Tennessee runs a line of black shale which contains gas. Thousands of feet thick in places, the Devonian shale (350 million years old) is a true shale, unlike the so-called shales of Colorado–Utah. The Appalachian shale contains gas in concentrations of between 24,000 and 36,000 BTU per metric ton, and the total resource has been estimated to be as large as 522 quads.[14] Again, the gas is difficult to obtain. Sometimes fracturing, which is injection of large amounts of water into the well to crack the shale, followed by injection of large amounts of sand or refractory materials to prop open the channels created by the water, is required to stimulate the flow of gas. One company estimated in 1977 that at a minimum it would need to sell gas at a price of almost $3.00 per million BTU; the price of gas now averages $3.00 per million BTU to residential customers.

4.3.3. Gas in Tight Sands

Sands as hard as concrete hold gas in the Rocky Mountains. The sands can be fractured to stimulate the flow of gas, but much more sand and water are required than for Devonian shale fracturing. Nonetheless, the higher prices once allowed on the intrastate market (approximately $3.00 per million BTU) have stimulated hundreds of wells in the tight sands that are producing today. Tight sands could ultimately produce 800 quads.[9]

4.3.4. Gas in Coal Seams

Another unconventional source of gas is coal seams. Every canary in a coal mine knows about gassy coal, the cause of mine explosions. In fact, the equivalent of one day's supply of oil for the U.S.—the BTU content of about 18 million barrels of oil—is vented to the atmosphere each year from Appalachian underground coal mines to reduce the hazard of explosion. One method of tapping this gas has been tested at an Eastern Associated Coal Corporation mine in West Virginia. The method involved sinking the mine shaft several years before the coal was to be mined. Over a three year period, methane in the equivalent of 24,000 tons of coal per year was collected and sold to a nearby pipeline. Many strip mines in Appalachia produce less energy than this per year. Reportedly, this rate of production can pay for the expense of sinking the shaft early, especially as gas prices increase. While the resource of more than 800 quads estimated to be recoverable in this manner certainly seems optimistic (and small in terms of the total available from the coal), much gas should be forthcoming from the Eastern coal fields as the value of gas rises. That it will can almost be assured because of the safety convenience inherent in the utilization of the hazardous gases.

4.4. Policy Implications

The husbandry of the precious hydrocarbon resources of the earth seems almost totally married to price, for better or for worse. If posterity has any oil or gas, it will be because economics dictated that the maximum sustainable production rate was not exceeded; that where substitutes such as coal could be used safely, consumers switched to the more abundant fuels; and that advanced recovery techniques such as tertiary crude recovery and Devonian shale fracturing were developed. Thus, while husbandry, fuel switching, and enhanced recovery are crucial fuel conservation policies, price dictates the possible.

What can we conclude about price from the foregoing? That the cost of oil and gas production is unlikely to decline or even hold steady but will probably increase at a rate which will result in a price that will more than double by the year 2000. If in fact the price doubles, then the substitution of solid fuels will be more attractive, especially if effective environmental controls are available. Synthetic fuels from coal or biomass would also be more nearly competitive. Synfuels could remain unattractive, however, if a large quantity of gas is made available at $6.00 per million BTU. Certainly, conventional natural gas is environmentally preferable to synthetics from coal or oil shale because of the vast human and ecological damage large scale use of those two sources would cause. Gas from unconventional sources will cost more than conventionally produced gas, perhaps even a factor of two to four more. In any case, these higher prices need not mean social upheaval. Even at the highest price mentioned, gas would still be less expensive than today's electricity. Gas should be the fuel on which we rely most heavily as we make the transition to renewables.

We must reiterate that higher energy prices present the challenge of achieving desired amenity levels with less energy by substituting less-expensive capital and ingenuity. In this sense it is well that economics dictates the rules, for time, capital, and labor are also scarce resources which must be allocated optimally. A chief concern is that all costs of production, environment, human health, etc. be internalized. All costs, insofar as possible, must be included in the price of energy.

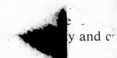
y and c'

Coal prices at present are dropping, and current estimates are that they
will remain stable or drop further for the rest of the century.

—SHELDON NOVICK
The Careless Atom (1969)

*"Oh Daddy won't you take me back to
Muhlenburg County
Down by the Green River where Paradise lay?"
"Well, I'm sorry my son but you're too
late in asking
Mr. Peabody's coal train has hauled it away."*

—JOHN PRINE
Paradise (1971)

We have underestimated the task of coal, and overstated the art.

—AGRICOLA
De Re Metallica (1526)

Chapter Five

Synthetic Energy Carriers—The Next Best Thing?

5.1. Introduction

Alternate energy sources are as numerous as lobby groups in
Washington. Although the implicit meaning of alternate is "the next best
thing," it is not clear which energy carrier, by which we mean the form of
energy delivered to final demand, is the best to replace petroleum. We are
not even certain whether it should be one or a combination of carriers, but it
is clear that future sources of energy should be as cheap as possible,
renewable (if possible), reliable, and should carry the lowest external costs.
While biomass and solar systems can plausibly meet these criteria, coal-
derived energy forms are more troublesome. Nuclear power, though it is
superficially clean, may be unacceptable for reasons other than price.

Crude oil is our major *source* of energy, but it is converted variously into gasoline, diesel, fuel oil, heavy oil, other refined products, and electricity for use by consumers. Alternate energy carriers include synthetic liquid and gaseous fuels, electricity, and direct energy carriers such as solid coal, biomass, and heat from the direct use of sunshine. We have reached a point where it is no longer possible to say, "I favor solar," or, "I'm a synfuel man," or "Coal is king," if one wants to be understood. Since synthetic liquid fuels, electricity, or direct heat can be made from any energy source, and since the costs of deriving these energy carriers from solar energy, coal, oil shale, or uranium can vary greatly, it is important to be specific. To speak of alternative energy sources in terms of broad categories can only be useful in masking ignorance or for effecting an emotional response while obscuring the real issues behind a choice of energy sources.

In this chapter we examine the production of electricity, synthetic fuels, and direct energy use for the purpose of identifying cost and resource constraints. This, coupled with the discussion of environmental problems which follows in the next chapter, will provide a context for evaluating energy conservation strategies.

5.2. Electricity

In Part I of this book we examined the scenario which would have us using exponentially increasing amounts of electricity, chiefly nuclear generated electricity, as the energy of the future. We reflected on a basic question, popularized by Amory Lovins, which is, "Is electricity really appropriate to our needs?" We intend to meet this question in Chapter Seven. But first we turn to a question which must be asked in parallel if not in advance of the question of appropriateness, and this is, "How much will electric energy cost?" As with the other energy forms, we present electric energy costs in terms of dollars per million BTU. Please be cautioned, however, that some BTU are more valuable than others, that high grade energy carriers such as electricity can be used more efficiently and diversely than lesser grade forms, and thus the costs of delivering energy services are not always directly comparable. We return to this matter in Chapter Seven under the heading of thermodynamics.

5.2.1. The Economic Cost of Nuclear Electricity

In any discussion of energy costs, costs may be broken down into the components of capital equipment, operating and maintenance (O & M), fuel (if applicable), and transmission or distribution costs. The cost of using oil is dominated by the fuel cost to such an extent that the cost of capital and O & M can be ignored somewhat in discussions of generation policy. The opposite is true for nuclear power, where the cost of capital, insofar as generation is concerned, is the dominating factor. In terms of delivered electric energy costs, however, the cost of manufacturing the electricity is frequently less than half the delivered cost. It is the transmission and distribution of the electricity which accounts for the balance.

The cost of power plant capital is enormous. A typical 1000 megawatt nuclear plant costs at least $1 billion ($1000 per kilowatt). Adding the cost of borrowing money, taxes, insurance, and so on, and levelizing (annualizing) these costs over the life of the facility results in a cost of about 1.9 cents per kilowatt hour.

O & M and nuclear fuel add an additional 0.2 cents per kilowatt hour. Combining all generation costs with transmission and distribution thus totals more than 4.5 cents per kilowatt hour, or about $13.20 per million BTU.*

Major questions arise at this point. Will depletion of rich uranium ores drive up the cost of nuclear fuel and, as a result, the price of electricity? Will this situation foreclose the nuclear option unless we rapidly develop a breeder reactor?

We confess that we cannot add much to the debate over the size of the U.S. uranium resource. The magnitude of this resource, of course, depends on how much one is willing to pay for the uranium. Prices currently run about $40 per pound for ore that is .2 percent fissile uranium. The U.S. Department of Energy has conservatively estimated that 4.4 million tons of such uranium ore are available in the U.S. for less than $120 per pound.

A typical 1000 megawatt pressurized water reactor (PWR) requires 143 tons of uranium annually, plus 242 tons initially. So, 4.4 million tons of uranium ore could supply 700 reactors for 30 years without recycling uranium. Recycling the unburned uranium in "spent" reactor fuel, fuel that has become poisoned by the collection of radioactive waste products so that a chain reaction cannot be sustained would extend the uranium resource to supply 1100 PWRs for 30 years.† For comparison, recall that only 70 reactors presently operate in the U.S., and even the most enthusiastic forecasts of the use of nuclear energy have never called for more than 1000 reactors.

---

*Coal-electric generation costs vary from region to region, but are roughly comparable to nuclear-electric costs on the average.

†This calculation ignores the fissile plutonium, bred in every power reactor, which *could* be recycled for fueling power plants.

The increase in total electric costs as a result of a $10 per pound increase in the price of uranium ore would be $0.0008 per kilowatt hour. If all uranium increased to $120 per pound, the total increase in the price of electricity would be only 15 percent.[1-3]

Thus we may ask, "Do we really need a crash program to develop a breeder?" Breeder advocates answer yes, arguing that the requisite fuel for beginning a plutonium or breeder cycle would be exhausted otherwise.[4] Another answer may be made to the question, however.

The present generation of reactors are to uranium what two-ton cars are to gasoline: guzzlers. In fact, according to one of the patent holders, the light water reactor (LWR) system was never intended for central station power generation. It was intended instead for submarine propulsion where inefficiency did not matter so much.[5] The adaptation of LWRs to commercial power generation was an historical accident due in part to the failure and slow development of breeder reactor designs. The development difficulties of breeders, it seems, were greatly underestimated.

An alternative midway between breeders and uranium guzzlers is the advanced converter reactor (ACR). In one of two ACR designs,* the heavy water reactor (HWR), the heavier, less common isotope of hydrogen serves as the moderator. In the HWR, water is composed of deuterium and oxygen. Because deuterium absorbs neutrons much less frequently than common hydrogen, the neutron economy in a reactor moderated by heavy water is greater and a breeding ratio of .6 can be obtained. But with recycling of the spent fuel, the breeding ratio can be pushed almost to one, the point at which as much fuel is created as is consumed. Harold Feiveson argues[6] that a combination of uranium recycling and ACR utilization could provide as much power over the next century as a liquid metal fast breeder program started now, and at comparable costs. This combination would not jeopardize the possibility of future reliance on a breeder reactor cycle, but would allow deferral of the decision to commit ourselves to plutonium for one hundred years.

Secondly, certain proponents of fusion argue that a special fusion device is possible which would generate copious quantities of neutrons which could breed fissile fuel starting with U-238 or Th-232. Further, a greater supply of fertile thorium than that of uranium exists, and from this resource a breeder reactor cycle could be built.[7]

Do we urgently need a breeder reactor? If we were to build 1000 reactors in the next ten years, the answer might be a qualified yes. But it seems highly unlikely that so many reactors will or even need to be built

---

*The second type of ACR is helium cooled, and is known as a High Temperature Gas-Cooled Reactor, or HTGR.

over the next two or three decades. It is unlikely even that such a level of demand for electricity would so seriously affect the uranium resource that the breeder would be needed to hold down electricity prices. The central issue of demand for electricity is one to which we will return in Chapter Seven.

Basic research as opposed to development and demonstration on breeder reactors, and, in particular, thorium cycle systems may be desirable, nevertheless, because we simply do not have as many options in hand as we would like. The reduced rate of growth in demand for electricity coupled with the *de facto* moratorium on nuclear development in the U.S. offers the opportunity to give further consideration to improved reactor designs and technology.

5.2.2. The Economic Cost of Coal-Fired Electricity

Coal, we are told, is the fuel of the future. We are urged to reduce the restrictions on its use. But the fact that coal use is destructive to the land, water, and human health, and could cause major changes in world climate is a good reason not to become excited over any nuclear moratorium strategy which substitutes coal-fired central station electric power generation for nuclear power. We discuss the details of the "curse of coal" in Chapter Six.

The cost of making electricity from coal is not much different from that of nuclear power. Capital costs may total only $700 to $800 million per gigawatt plant, but fuel costs at about one cent per kilowatt hour are substantial.[8] The cost of transmission and distribution again comprises about half the delivered energy cost of 4 to 4.5 cents per kilowatt hour.[9]

What will happen to the cost of steam coal in the future? The price of coal in the past has roughly paralleled that of oil. When the price of oil increased, so did coal, albeit for different reasons. It is obvious that the future of coal prices will be closely tied to demand. The present production level of about 700 million tons per year—more than half from Kentucky, Pennsylvania, and West Virginia—is about the same as in 1920.[10] A number of large scale synthetic fuels plants would of course drive up demand and demand would drive up the price of coal. But a four percent annual real rate of increase in the cost of steam coal over the next twenty years (a doubling) would add only 25 percent to the cost of delivered electricity.

A debate has raged for years over coal resource estimates, but it has been focused less on how much coal exists than where low sulfur coal deposits may be recovered. The issue presented was whether coal should come from the west, where the percentage of sulfur by weight is low, or from the east and be used in conjunction with sulfur removal. As we explain in the next chapter, sulfur emissions are creating such extensive health and environmental problems that the issue should be superfluous. Virtually all coal combustion should be coupled with sulfur removal. In any case, if the lower calorific value of western coals is taken into account, one finds that the difference in sulfur associated with eastern and western coal is significantly reduced.

A concurrent debate involves the issue of whether coal should come from surface or underground mines. Coal generally costs electric utilities slightly more than $1 per million BTU, or $20 per ton, figures which assume that a ton contains more than 20 million BTU.*[11] Strip mined coal historically costs about $1.50 per ton less than deep mined coal.[12] The Federal Coal Surface Mining Regulation and Reclamation Act of 1977 added about $0.10 per million BTU to the cost of strip mined coal, or about $0.0003 per kilowatt hour of delivered electricity.[13] As we discuss, the act is far from a complete control of strip mining abuses, but still its benefits far outweigh the small costs it imposes on the price of energy.

The coal resource is so vast that the price of coal should not be affected by resource constraints for at least a century.† The resource is far from as large as has been widely claimed, however. The coal resource, defined as coal that lies within 3000 feet of the surface (90 percent actually lies within 1000 feet), totals 1.6 trillion tons, enough, as they say, to supply the U.S. with coal for 500 years at the current consumption rate. Not all of this coal is recoverable, however. Present mining technology recovers less than half the coal in place. While strip mining is more efficient strictly in terms of coal extraction, it can recover only that 15 percent of American coal which lies within 150 feet of the surface. Thus, more than 85 percent of the total coal resource will have to be deep mined if it is to be obtained. Room and pillar underground mining methods recover less than 50 percent of the coal in a given seam, but longwall underground methods would recover 80 percent.[10,12,13]

Longwall underground mining, in its simplest terms, is a method of underground mining which replaces coal pillars with mechanical lifts for roof support. When the coal is mined out, the roof supports are advanced, and the ceiling collapses. This method not only improves recovery rates, but it reduces the impact of surface subsidence. Most importantly, it may be ten times safer for the miner than room and pillar mining.

Promoting longwall mining will probably require government stimuli because the initial cost of the equipment is very high, perhaps three times as

---

*Low sulfur coal may sell at a premium, ranging from an additional $5 to $20 per ton, or $0.20 to $0.80 per million BTU.

†This is not to say there will be no production constraints.

great as for conventional equipment. Longwall mining is feasible, however, because 90 percent of European coal is mined in this way. Currently, though, less than five percent of American coal is longwall mined. The reward to the mine operators could be substantial since the method has a labor productivity potential ten times greater than conventional mining. Further, the value of the coal in place is increased by more than fifty percent by the higher recovery rate. But the high investment required is discouraged by the uncertainty of the coal market. Government loan guarantees or tax credits for longwall underground mining investments would promote the technology. In so doing, the government could help ameliorate the national disgrace that exists in the death and destruction caused by coal mining and provide in the bargain an important energy conservation tool.[12]

Industrial cogeneration, the concurrent generation of steam for both industrial process applications and electrical generation, is one important option open for coal utilization (see Chapter Ten). Whereas large-scale, central station coal-fired electrical generation offers little cost advantage over nuclear power, the use of coal-fired cogeneration in smaller size plants can reduce both the fuel and capital costs to steam and electricity users.* In cogeneration, capital costs may be shared between steam and electric consumers, both of whom reap the benefits of vastly improved thermodynamics. The cost of steam generated by cogeneration may be two to three dollars per million BTU (two to three dollars per 1000 pounds of steam) cheaper than if the steam were generated by burning gas or oil. Cogenerated electricity generally costs between one to one and a half cents per kilowatt hour ($3.00 to $4.50 per million BTU) less than central station generated electric power.[9]

While there are limited prospects for reducing the electric generating costs using nuclear fuel, the potential for coupling industrial steam generation with electric power generation in small, coal-fired industrial plants promises slightly reduced electric power costs. The institutional problems of implementing industrial cogeneration are discussed in Chapter Ten.

5.2.3. The Economic Cost of Solar Electricity

There are three major ways of making electricity using solar energy: solar photovoltaic conversion, biomass combustion for steam-electric generation, and "tower-power" or central receiver steam–electric power generation.

---

*Cogeneration is generally not practical at large central station power plants.

The use of solar photovoltaic silicon or gallium arsenide cells for the direct conversion of sunlight to electricity is one of those systems which captures the imagination with its sheer elegance. The seeming simplicity of generating power from roof panels generates an enthusiasm that is marred only by a few concerns. The cost at present is twenty times greater than the going price of electricity, and its storage and/or integration in utility systems are unsolved problems. These problems, however, are subject to amelioration. Rumors of secret breakthroughs in the manufacture of silicon cells may be valid, and the cost of photovoltaics may drop sharply within a decade. In the meantime, experimentation with the integration of photovoltaic operation with utilities should be practiced to gain the knowledge necessary to apply solar photovoltaics when and if the cost breakthrough comes.

The use of biomass for steam-electric generation or simply for industrial process steam generation could make use of significant existing quantities of municipal solid waste, forestry, and crop residues, and perhaps someday, energy crops. However, energy crops and the use of forestry and agriculture residues may represent serious environmental threats to soil productivity and water pollution (see Chapter Six). The total potential biomass resource is large and, if properly managed, could contribute perhaps 15 quads by the end of the century in addition to the 1–2 quads of wood waste currently used by the forest products industry.*[15] Of course we will be fortunate if half of this figure is achieved by the end of the century.

The cost of using solid biomass fuels in electrical generation is comparable to that for burning coal, though total costs are reduced when otherwise valueless fuels are used. A cogeneration plant in Eugene, Oregon (described more fully in Chapter Ten) burns wood waste costing only $0.33 per million BTU, and produces electricity for only 1.8 cents per kilowatt hour.[16]

Gasified biomass fuel costs more than solid biomass fuel because of the capital equipment required for gasification and the inefficiency inherent in conversion. But gas turbine electric generators are far cheaper than steam-cycle systems (gas systems cost only about $200 per kilowatt, solid fuel systems $1000 to $1500) and can deliver electricity for three to four cents per kilowatt hour. Both types of plants are easily adaptable to cogeneration and can deliver steam to industries for about $3.00 per million BTU, a price competitive with steam generated by burning natural gas or coal.[9]

---

*Including 5 to 10 quads of wood, one to five quads of crop residues, 1 quad from energy farms, 1 quad from municipal solid waste, perhaps none from grain or sugar crops.

Tower-power involves positioning mirrors in rows or rings on a desert, and directing the reflected light to a central receiving station, a tower, where the heat of extremely concentrated light boils water for steam–electric generation. Due to the prodigious quantities of materials required, the cost will surely run to 10 cents per kilowatt hour ($30.00 per million BTU), with little prospect for reduction.[17]

So it appears that solar energy could potentially provide a significant source of electric power at competitive prices. Photovoltaic electric generation, however, will not be economically available for at least a decade. Biomass–electric generation, especially coupled with industrial process steam generation (cogeneration), could make use of several quads of available wood waste immediately.

5.2.4. Inflation and Electric Energy

During the second half of the decade of the 1970s, the rate of cost increase in the utility construction sector averaged two to three percentage points above the average rate of inflation. This real price increase in both the materials and labor which go into electric power plants drastically altered the nation's perception of electricity as the energy carrier of the future. Systems which inherently carry large capital and labor costs—coal, nuclear and solar—became less competitive with systems which require lesser capital investment. So the energy produced by these systems did not become cheaper than oil and gas despite enormous petroleum price increases. It is unlikely that the trend of real price increases for capital equipment will change in the foreseeable future, so it is unlikely that electricity will become significantly more competitive with other energy carriers.

Likewise, enormous amounts of capital are required for synthetic fuels plants. The inherent cost of capital partially explains why synfuels have not become competitive and are unlikely to become so in the near future.

5.3. Synthetic Fuels

Synthetic fuels pose a grave issue. It is an issue which rouses our deepest fears and hopes and raises many questions. Should government build and operate synthetic fuels facilities, or should such development be left to private industry and the market? Should the water rights of farmers in western states be appropriated for synthetic fuels producers? Should Appalachian residents endure more strip mining and see coal refineries built in order that others may enjoy their automobiles? Must miners die of black lung so that automakers may not have to redesign their cars?

Ultimately, we will have to convert our fuel economy to renewable synthetic fuels. We can make this conversion logically, in a stepwise, economical and environmentally acceptable manner over several decades, or we can attempt a moonshot type crash effort to produce synfuels. We may impose environmental suffering on some to avert energy shortage for others. Or we may avert both energy shortage and environmental suffering. Or we may wreak havoc with the environment and still fail to produce usable synthetic fuel.

5.3.1. The Feasibility and Economic Cost of Synthetic Fuels Production

Synthetic fuels could be produced in a variety of forms, from gasoline to alcohol, from solvent refined solid coal to substitute natural gas. Feedstocks or raw materials may include coal, oil shale, wood, crops, and garbage. Synfuels can come from anything, in fact, that can contribute the elements hydrogen and carbon.*

Hydrogen and carbon are building blocks from which an enormous variety of organic compounds can be synthesized because of the almost unique valence of carbon (carbon can form four chemical bonds per atom). Energy is released when the bonds between carbon and hydrogen are broken by oxidation, thus forming heat, carbon dioxide, and water. The physical state of hydrocarbon compounds (their solid, liquid, or gaseous state) depends, *caeteris paribus,* on the ratio of hydrogen to carbon atoms in each molecule. Methane (natural) gas, with a formula of CH_4, has the highest such ratio and is one of the most elegant fuels. It burns readily, cleanly, and cheaply, and is transported easily in pipelines. Gasoline, refined from crude oil, has an average formula of $C_8 H_{18}$, while bituminous coal has an average formula of $[C_{13}H_{10}O]_n$, and may have an average molecular weight of 1000 to 3000. Coal is a vastly complex substance with not only many chains and rings in its chemical structure, but also many inorganic elements and chemicals including much water and frequently a high percentage of sulfur. As a solid it is difficult to transport and use, and its combustion yields a number of noxious substances.

The technology of synthesizing liquid and gaseous fuels from the building blocks of hydrogen and carbon, however, may make solid hydrocarbon fuels usable in important ways. But it should not be a foregone conclusion that the optimum solution of our imported oil dilemma is the direct replacement of gasoline, diesel, and fuel oil from crude oil with

---

*Hydrogen alone makes an almost ideal gaseous fuel: it reacts with oxygen to form water and heat.

these same products made from domestic solids. The problem is as complex as coal itself.[18]

5.3.1.1. Synfuels from Coal

Solid coal may be refined to a relatively clean solid fuel by use of solvents. Coal may be gasified to produce a boiler or furnace fuel and thus be a substitute for natural gas. And coal may be converted to liquid fuels by four methods: (1) hydroliquefaction (the liquefaction of coal by direct reaction with hydrogen gas in the presence of catalysts); (2) solvent extraction (the noncatalytic reaction of coal with hydrogen donated by a liquid solvent); (3) pyrolysis (the heating or destructive distillation of coal in the absence of air); and (4) catalytic synthesis (which involves two steps, gasification, and synthesis of the gaseous products to make liquids). The first two methods are referred to as direct liquefaction, and the last one, widely known as Fischer–Tropsch synthesis, as indirect liquefaction. Pyrolysis converts a greater percentage of the feed coal to char than to gas and oil and is thus the least interesting of the four.

Coal can be gasified by reacting its carbon atoms with steam. This reaction requires heat which is usually provided by burning some of the coal, and yields carbon monoxide and hydrogen. Methane (CH_4), the principal constituent of natural gas, may be produced by reacting hydrogen with carbon monoxide in a three-to-one ratio. In this case the additional hydrogen is produced by reacting carbon monoxide with water. Producing methane in this manner is called methanation, and the complete process is called simple gasification.

A second gasification method, called hydrogasification, results in the direct reaction of hydrogen with coal. Instead of heating coal with steam to produce carbon monoxide and hydrogen, from which methane is generated, hydrogen is first produced in a separate vessel and is combined with coal to create methane and char.

The Fischer–Tropsch coal liquefaction process begins with simple gasification, then proceeds with it only to the extent that carbon monoxide and hydrogen are produced. Methanation is not performed. Instead, iron catalysts are employed to react the CO and H_2 to form a wide range of liquid hydrocarbons including gasoline, fuel oil, and both heavier and lighter products. But in this process two-thirds of the feed coal is lost.

Methanol (methyl, or wood, alcohol) can also be synthesized in the Fischer–Tropsch process by the use of different catalysts. Copper–zinc–chromium combinations, for example, will favor the CO and H_2 reaction and thus result in CH_3OH (methanol).[18]

Alcohols are fine fuels straight or blended with gasoline. Gasohol is a blend of 10 percent ethanol [$CH_5(OH)_2$] (also known as grain, or ethyl, alcohol), and 90 percent gasoline. Gasohol may be made from methanol but blends are limited to only five percent methanol.* Methanol will, in concentrations greater than five percent, damage gaskets and seals in gasoline engines. Diesel engines cannot use methanol without major modifications.

A substitute crude oil, for which products like gasoline or distillate fuel may be refined, can be made from coal by combining coal with a solvent rich in hydrogen. Solvent extraction, as this process is called, requires no catalysts which themselves are usually imported metals. Some solid coal is left unliquefied, but this coal can be reacted with steam to produce the hydrogen necessary for the process. A solvent less rich in hydrogen, as mentioned earlier in this discussion, produces clean but solid coal.

Coal can also be liquefied directly by first pulverizing and mixing it as a slurry with coal-derived oil, and then mixing the coal slurry with gaseous hydrogen in a reactor containing a catalyst such as cobalt molybdate, or tungsten. The product is a heavy syncrude. Syncrude can replace crude oil in a conventional refinery.

Each method, whether gasification, direct liquefaction, or indirect liquefaction, has advantages and disadvantages. One common disadvantage is lack of commercial development. Even the most commercially advanced process, Fischer–Tropsch, is not yet viable despite oil price increases. Fischer–Tropsch seems to be popular in the media, but there is only one commercially operating plant in the world, at Sasolburg, South Africa. This plant produces only 10,000 barrels of liquids per day, much of which is in the form of products which would not match U.S. markets very well.

Two factors which make the South African Fischer–Tropsch coal conversion operation worthwhile to South Africans are (1) their nation's inability to buy oil at any price because embargoes were imposed on them in protest of apartheid, and (2) because coal mining labor is available at what amounts to slave wages. Clearly, Fischer–Tropsch provides a technically feasible means for producing synfuels. Just as clearly, however, the economics of Fischer–Tropsch are unfavorable.

How much would Fischer–Tropsch gasoline cost? How much would any synthetic fuel derived from coal cost? Answers are uncertain for many reasons, but the range seems to be the equivalent of $2.00 to $6.00 per gallon of gasoline ($16 to $48 per million BTU).

Some examples of the costs of the various processes should substantiate this conclusion. First, there is the capital equipment consisting of

---

*The gasohol blends used today are made with ethanol distilled from corn.

coal washers and crushers, gasification reactors, oxygen plants,* shift reactors,* acid gas and sulfur plants, reactors, etc. A plant of just about any kind for making 30,000 barrels of liquid fuels per day would cost anywhere from one to three billion dollars. Judging from cost overruns on military equipment, power plants, large construction projects, and the like, it would not be unreasonable to expect a doubling or tripling of this cost. Even with the lower cost, Fischer–Tropsch gasoline (delivered) would cost $16.00 per million BTU, or $2.00 per gallon.

Half of this cost is the coal feedstock which is usually assumed to sell for $1.00 per million BTU. Whether this price will hold in real terms in a market in which the demand for coal would double, as it is assumed to do in any coal synfuel scenario, is a tenuous bet.

One-fourth of the total cost goes for operation and maintenance costs, a large part of which is labor. Without invoking slavery, we may expect this cost to increase faster than the rate of inflation.

The prospect of an economic "breakthrough" in synthetic fuels production seems unlikely. The basic demands for expensive materials, the inefficiency of fuel conversion, and the great risk involved for any investor are sure to place synthetic liquid fuel production from coal far from the top of the list of next best things.

Coal gasification, being simpler, is less expensive. At the same time, the market in which coal gas would compete is more exacting than the gasoline market. Thus natural gas relative to synthetic gas is a better bargain than is gasoline from crude oil relative to synthetic gasoline. Again, the capital, feedstock, and maintenance costs are high. Gas from coal would cost $6.00 to $10.00 per million BTU compared with $3.00 to $4.00 for natural gas and $5.00 to $6.00 for unconventional natural gas.[15]

5.3.1.2. Synfuels from Oil Shale

Kerogen, a bituminous substance contained in oil shale, yields oil when heated. Oil from shale may be substituted for crude oil. The resource in the Colorado–Utah area is vast, perhaps as great as 600 billion barrels. Technical and economic problems which may keep this resource in the ground, however, are (1) the fact that the kerogen is contained in the pores of the shale rock itself, and enormous quantities of rock must be processed to obtain a small quantity of oil, and (2) much water is required in the

---

*In some processes.

process of retorting* oil from shale. A ton of rock will yield 20–40 gallons of oil, or between three and six million BTU. By comparison, most coal mined in the U.S. today contains 20–25 million BTU per ton. Water requirements for oil shale retorting may total five barrels per barrel of oil. Even ignoring the environmental prospect of disposing of one ton of crushed shale for every twenty gallons of oil produced, and that of using prodigious quantities of water in a region where water is a bitterly contested resource, assessing the cost of producing petroleum in this way would seem prohibitive. Yet, estimates falsely place oil shale gasoline at $10–$15 per million BTU, or $1.20 to $1.90 per gallon. We expect that fuel from shale, if produced, will cost far more.

5.3.1.3. Synfuels from Biomass

Gasohol is a popular fuel. Americans are willing to pay a premium (5 cents per gallon extra) for it because it provides five percent better gasoline mileage and because it is home grown. Improved gasoline mileage results because the ten percent ethanol in gasohol upgrades the octane rating of regular gasoline, and also because it burns leaner (that is, with less fuel per unit of air), since ethanol contains fewer BTU per gallon.

The issue of whether gasohol requires more energy than it saves appears to depend on the fuel used in its manufacture. The manufacturing process involves hydrolyzing biomass such as corn to sugars and then fermenting the sugars. This process, including removal of water and impurities, costs about $1.50 per gallon of ethanol. But gasohol is made competitive with gasoline by a major government incentive, the removal of excise taxes of four cents or more per gallon of gasohol. This federal subsidy is equal to $0.40 per gallon of ethanol, or $16.80 per barrel.

There are several objections to gasohol. Making fuel from corn may deprive the hungry of badly needed food. If gasohol is made in a distillery using premium fuel, more petroleum is burned in making gasohol than it replaces. And making lots of energy from biomass could lead to an environmental Armageddon. Leaving Armageddon to the next chapter, let us examine the first two issues.

It is true that there is a real danger that American buying power will put corn in the tank rather than in the bellies of the hungry. But at the current level of ethanol fuel production, this is not likely. If anything, by supplementing farm income, limited gasohol production from grain should

---

*Retorting is synonomous with destructive distillation, that is, the heating of (in this case) oil shale in the absence of air.

stabilize farm markets. A problem would arise, however, if America pursued some "Manhattan project" sort of energy-from-food scheme.

In the long run it will be biomass which provides a source of hydrocarbons for fuel or any other purpose.* To plan carefully to make a transition to gaseous or liquid fuels from renewable biomass while sidestepping the solid fossil fuels could be the wisest course for America to take. Consider the following scenario.

America has a natural gas pipeline system in place. The pipeline system can serve 75 percent of the population. The country also has the capability of producing 5–10 quads of biomass energy presently and 10–15 quads by the year 2000. A workable strategy could be to shift energy-efficient homes and factories to natural gas, using at first domestic conventional and unconventional gas supplies, and then switching gradually to gas made from biomass. The use of only perennial grasses and like sources instead of food and crop residues for hydrocarbons would reduce erosion and other adverse impacts. As natural gas becomes more scarce and expensive, gas from biomass plants could be added all along the pipeline system, thus decentralizing production. This arrangement would carry both environmental and strategic advantages. Oil replaced by gas could be used more frugally in the transportation sector. The cost will not be unreasonable. Indeed, renewable energy supplies could actually set a ceiling on energy prices, because resource depletion need never be a factor in driving up price.[14]

5.3.2. Summary

The data presented in the above narrative are summarized in a way that is more readily comparable in Figure 7.2 (p. 139). Note how the projected costs for the various energy carriers compare. And note how electricity produced by cogeneration, and the direct use of fossil fuels, is cheaper.

5.4. Direct Use of Solid Fuels

Investment in the direct combustion of coal in industry might be a strategy superior to either synthetic fuels production or electric generation. Industrial boilers can be built now using fluidized bed combustion which minimizes sulfur emission, suppresses particulates, and reduces the emission of other trace elements. The cost of generating steam from coal is

---

*Methanol, ethanol, ammonia, methane, hydrogen and solid fuels may be produced from biomass.

about $3.00 per million BTU. Steam accounts for about 25% of industrial energy end use. But direct application of process heat accounts for more than 50 percent. Fluidized bed combustors can furnish direct heat at a cost roughly one-half that for synthetic gas to do the same job. Displaced oil (industry uses 2.5 million barrels of oil per day) could be shifted to the transportation sector. Although much of this oil is of low quality, refinery retrofits could be made to upgrade the oil for higher uses.

5.5. Direct Use of Solar Energy

Despite all common wisdom that solar energy turns into a pumpkin at sundown, there indeed seems to be a Cinderella among the ugly sisters of energy supply.

It is a time-honored practice to relegate all technologies or options which one wants to ignore, deemphasize, or discount to another category. In this tradition and for a variety of reasons we will stigmatize new hydroelectric sources, solar central station electric (tower-power), ocean–thermal energy systems in our solar discussion, and solar space satellites. But certainly not the direct use of solar energy. Indeed, of all energy supply options, passive solar energy technologies are not only the cheapest economically, but the most benign environmentally.

5.5.1. Industrial Solar Applications

About 40 percent of industry's energy use is in the form of hot water or steam between the temperature of 170° and 570°F, a range that solar flat-plate collectors and the direct combustion of biomass are capable of supplying. Industries have already started applying solar water heating to their processes. Campbell Soup in Sacramento, California, for instance, washes cans with 190°F water heated with roof collectors; Anheuser-Busch in Jacksonville, Florida, washes beer bottles with water heated in roof-top collector tubes; a textile plant near Atlanta, Georgia, *Shannandoah,* will in the early 1980s obtain 1000 pounds of steam per hour from solar collectors.

Industry now pays somewhere in the vicinity of $2.00 to $3.00 per 1000 pounds of steam (1000 pounds of steam roughly equals one million BTU). Solar collectors can supply steam for a cost of $3.00 to $15.00 per million BTU. Industrial application of solar energy is rapidly becoming competitive, particularly in industries which have maintenance crews already available to service solar systems.[14]

The effect of solar energy in industry will not be to lower present energy prices, but rather to place a ceiling on future costs.

5.5.2. Solar Energy in Buildings

Solar passive and active space conditioning begin to look attractive at energy prices of about $5.00 and $15.00 per million BTU respectively. Much of the benefit of solar passive design derives from concomitant large investments in insulation, however, and is therefore difficult to distinguish from straightforward conservation. Operationally, solar passive should not be distinguished from the other options that can be used in what is called energy-conscious design. But that is far from saying that solar is unimportant. A house incorporating energy-conscious design can realize a savings for its owner of more than 50 percent in space-conditioning fuel costs. Often it requires little more than building with lots of glass facing south to absorb heat in winter and with no glass on the north to prevent heat loss. It may include lots of insulation to keep summer heat out but winter heat in and devices such as insulating shutters, western and southern overhangs which block summer sun but utilize the low-angle winter sun. It would naturally include intelligent tree planting. Such homes may be heated almost entirely with passive design at a maximum marginal cost of about $20.00 per million BTU.

In active systems, leaks, snow accumulating on the collectors, dust gathering inside the collectors, high humidity in summer, and other small problems are frequently experienced, but are generally amenable to simple solutions.

The largest problem with active solar heating still is price, for $15.00 per million BTU is 50 percent higher than the average cost of electric resistance heating. Expensive electric heat pumps, though twice as efficient as resistance heaters, deliver heat at a marginal cost of about $11.00 per million BTU. There is no one number, of course, which reflects the cost of solar energy, since price is a function of factors of climate, house size, and so forth.[17]

5.6. Alternate Energy Carriers and Resource Husbandry

Even renewable fuels must be conserved. Images of deforestation and soil depletion (see Chapter Three) emphasize the need for conservation as much as do images of devastated coalfields. The choice of harvest or extraction technologies and energy conversion systems can have a profound effect on the availability of energy from both renewable and nonrenewable resources.

One scenario should serve to illustrate the importance of this issue. The Fischer–Tropsch process of converting coal to gasoline wastes 70 percent of the energy in coal. Assume that we would like to produce two million barrels of gasoline per day using Fischer–Tropsch plants. We would have to double coal production, and that would force an increase in mining of 700 million tons per year. If, however, a more advanced liquefaction process were chosen, one such as the H-Coal process, which is 60 percent efficient, only half as much additional coal would be required. On the other hand, if coal were used directly in industry and the oil displaced were shifted to transportation uses, then only one-third as much additional coal would have to be mined.

The implication of such a choice for the future of coal supply should be obvious. Assuming that four hundred billion tons of coal can be mined in the U.S., then the coal resource really could supply our present coal demand for 500 years. But if this resource is mined by underground room and pillar methods and converted to gasoline via Fischer–Tropsch to produce two million barrels of gasoline per day, then coal would last only 100 years. That assumes other uses of coal do not increase. Longwall mining could increase the useful supply of coal by fifty percent. Direct use of coal plus longwall mining would increase the life of the coal supply by more than three hundred years assuming a usage level for direct application of coal equivalent to two million barrels of oil per day.

Some might object that the depletion scenario is terminally pessimistic, that newer, more efficient coal conversion processes are forthcoming, that strip mining is more efficient than deep mining, and that unforeseen technical breakthroughs will aid us. But, even if we were not in the midst of a chaotic rush to produce synthetic gasoline by whatever means is available (i.e., Fischer–Tropsch synthesis), and even if strip mining really were feasible in the majority of our coal seams or even as efficient (it is actually less efficient than underground mining in Appalachia since it usually retrieves only the outcrop of coal and locks up most of the rest), we would be remiss not to pursue resource husbandry with all deliberate speed. To waste coal is to waste lives and nature, as well as energy.

5.7. Synthetic and Alternative Fuels: Policy Implications

Synthetic natural gas, at about $6.00 per million BTU, appears to be the cheapest synfuel. Directly liquefied coal will cost anywhere from $15.00 to $30.00 per million BTU after refining to gasoline or other suitable products. Indirectly liquefied coal, such as Fischer–Tropsch gasoline, will cost even more. Synfuels derived from oil shale most likely will be only slightly cheaper than synfuels from coal. Synfuels from biomass will also be costly but can be renewable. Electricity from new power plants costs $15.00 or more per million BTU, while natural gas, gasoline, and crude oil, cost $3.00, $10.00, and $5.00 per million BTU, respectively. It is clear that the only way for energy prices to go is up.

Only passive solar energy systems are significantly cheaper than most of the other supply alternatives, and passive solar energy is difficult to distinguish from energy conservation. The enormous potential for saving energy at costs of from $3.00 to $10.00 per million BTU should be obvious. The technical details of conservation are the subjects of Chapters 8, 9, and 10, in Part III.

Should we subsidize a synthetic fuels industry with billions of dollars of federal money? Such sudden heaping of money on the fire would dissipate our ability to cope with the energy problem. It is not necessary that we frantically dig up our countryside and turn our rivers and hills into gasoline. Conservation of energy, because it is cheaper, because it is quicker, because it is safer, and because it is more equitable, should help us avert the need for such a catastrophe. Produce we must, but our pace toward major synfuel production should be measured by real world experience from careful full scale testing.

Everything that lives resists. That which does not resist allows itself to be cut up piecemeal.

—CLEMENCEAU

There's a bit of strip-mined coal in every shirt.

—ANONYMOUS

Pave paradise, put up a parking lot with a pink hotel, a boutique, and a swinging hot spot.

—JONI MITCHELL
"Big Yellow Taxi" (1969)

Chapter Six

Conjunctures of Energy and Environment

6.1. Disasters

Energy-related environmental disasters, like pink hotels in Florida, are legion. Acute conjunctures of energy use and environment are deadly. It need only be recalled that an air temperature inversion over industrial Donora, Pennsylvania, in 1948, killed 20 and sickened 6000; that a December London smog in 1952 killed 4000 over five days; that heavy inversions in New York City killed 405 and 168, respectively, in 1963 and 1966 (the latter luckily occurred over Thanksgiving weekend when industrial activity was much reduced); that the oil tanker Torrey Canyon broke up in the English Channel in 1967 spilling 700,000 barrels of crude oil which killed thousands of seafowl, poisoned untold numbers of shellfish, and covered hundreds of miles of Britain's and Brittany's shorelines; that 1549 miners died in Honkeiko Colliery in Manchuria in 1942 in the world's worst single coal mine disaster; and that in February 1972, a small dam made of deep-mine spoil collapsed when spoil from a nearby strip mine slid into the reservoir adding the last stress the dam could take, and killed more than 100 people as a wall of water washed along Buffalo Creek in West Virginia. Of course, these are but a few of the worst examples. Now we have Ixtoc, the world's most recent record-size oil spill, and Tellico Dam, man's first intentional act of eradicating a species.

The chronic problems are even more devastating. Coal-fired generation of electricity causes between 2000 and 20,000 deaths each year with total health effects costing perhaps more than $2 billion annually.* Two million acres and 10,000 miles of streams in Appalachia have been devastated by coal strip mining, and air pollution damage to agriculture each year costs as much as $500 million. Automotive, industrial, and other sources of waste oil add twenty times as much oil to the sea as tanker accidents, and as many as 5000 species disappear from the Earth per decade, a rate 250 times higher than before man appeared.[1-6]

Finally there is a growing concern over the other global problems which may be caused by energy use. There is concern over the nuclear weapons proliferation and problems of nuclear power development. There is concern about the CO_2 problem associated with fossil fuel use and possible deforestation. Icecap melting and climatic shifts which may arise as much from particulates accumulating in the atmosphere as from greenhouse effect caused by CO_2 buildup are potential problems of tremendous importance which we may not fully understand for thirty years, at which time we may already begin to see effects.

It is worth emphasizing that energy consumers do not pay the full cost of their energy use. They have been subsidized by miners working in unsafe and unhealthy conditions and by fishermen whose livelihoods are damaged or destroyed when sea creatures are destroyed by toxic oil. They are subsidized by the inhabitants of Appalachia whose bad fortune it is to dwell near, but not own, coal resources (in strip-mined watersheds damaged from blasting, slides, and flooding can reach $4000 per household). They are subsidized also by each person whose health and/or property is damaged by air pollution and by the general diminution of the ecological health and aesthetic quality of the planet. The most widely proposed remedy for these problems has been to add their costs to utility and fuel bills. Such a remedy would internalize the external costs of energy production. However, not only are the costs and the benefits difficult to quantify, but philosophies differ on how internalization should be accomplished.

---

*Not counting the costs of Black Lung disease widely prevalent in coal miners. The public cost of attempting to compensate for Black Lung suffering totals $1 billion per year.

6.2. Control Controversy

Environmental law has evolved through complex debates on equity, utility, common law, free enterprise vs. market control, states' rights, benefits vs. costs, technical feasibility, and administrative practicality.

Common law, imported from England, was applied from the beginning of the industrial revolution to reconcile nuisance and tort with the utility of production. It was invoked, for example, to require piggeries to increase the height of their chimneys so the odor would be carried away. Nuisance law, it has been suggested, provided an inclusive if not concise history of industrialization.[7]

Common law was not adequate, unfortunately, to deal with the overwhelming quantity and diversity of health, property, and environmental insult generated by industrial production and consumption of industrial goods. Neither was the marketplace by definition able to deal with external costs. Not only is it counter to the spirit and incentives of free enterprise to pay more than one has to, but even if a business or industry were altruistic enough to pay for the nuisance and damage it caused, it often would be very near impossible to determine who was damaged and by how much. Science simply has been unable to connect, for example, a particular case of chronic heart and lung disease with a given air polluting factory. It would seem that all people, consumers and producers alike, were to blame. Hannah Arendt warned, however, that blaming everyone is the best way of saying that no one is to blame, and therefore avoiding punishment of the real culprits.

Common law in the U.S. first gave way to statutory environmental control with the Rivers and Harbors Act of 1899, otherwise known as the Refuse Act. The Refuse Act was a water pollution control law. It forbade the discharge of refuse into water from any ship in order to prevent refuse from washing into the nation's waters. It forebade the dumping of material on the banks of a body of water. Its main thrust was control of the source of pollution, as opposed to setting standards for water quality. The Refuse Act neglected to control sewage, unfortunately, but it did provide for fines for other polluters and established as well the framework for the requirement of discharge permits. Permits were not required, however, until 1970.[7]

This control of effluents from specific sources has coexisted with the Federal Water Pollution Control Act (FWPCA) which, when first passed in 1948, simply set water quality standards by specifying the amount of pollution which could exist in the nation's waters.

The FWPCA was amended five times before Congress gave it any teeth. In 1972, the philosophy of effluent control won out, with Congress setting limits on pollution discharge from point sources like sewage pipes, culverts, or any single, identifiable source. Nonpoint sources of pollution were virtually ignored, although these sources which include agriculture, forestry, and construction, may constitute at least half of our water pollution problem. The FWPCA required permits for point-source discharges (Section 402), provided fines for noncompliance (up to $10,000 per day—Section 309), and authorized huge expenditures to construct municipal sewage treatment facilities. Under Section 208, states were to be provided with funds to plan for control of nonpoint source pollution with the stipulation that any state failing to do so could lose its sewage treatment facility funding, a huge incentive since the federal government pays (under FWPCA authorization) 75 percent of the cost of these plants. Yet Section 208 planning, which essentially authorized land use planning, has only just begun. Finally, while not its primary thrust, the Act did set the goals of restoring and maintaining the chemical, physical, and biological integrity of the nation's waters, provided for the protection and propagation of fish, shellfish, and wildlife, and provided for recreation in and on the water by July 1, 1983. In other words, it set the goal of fishable and swimmable waters by 1983.

These goals have been much maligned for their ambiguity (and for a certain Orwellian taint). Some dispute the validity of indices for water quality such as quantity of dissolved oxygen, citing the variation and uncertainty in its measurement, particularly in estuaries, and their irrelevance to aesthetics.[8-10] Others insist that expenditures for effluent control should be based on their marginal utility. These argue that we cannot afford unlimited expenditures and thus should perform cost-benefit analyses to determine optimum investments. Still others, including politicians but mainly environmental lobbyists, argue that even an arbitrary standard that is enforceable is better than an amorphous set of standards which may be based on scientific exactitude. Their concern is getting the job of environmental quality control done, even if it is not done for exactly the right reasons.

The FWPCA of 1972, of course, reflects this attitude of expediency. Standards, like those in the act after which it was modeled, the Clean Air Act of 1970, are somewhat arbitrary. The basic question is, how does this approach work?

The Clean Air Act Amendments of 1970, too, had predecessors which embraced other philosophies of control. The Air Pollution Control Act of 1955 delegated the authority to the states. While the states' rights argument that concentration of power in the centralized federal government is not only unconstitutional but dangerous has appeal, reality severely imposes itself in such matters as air pollution. States have strong incentives not to mandate controls stricter than other states. To require expensive emissions controls or cleaner fuels would disadvantage a state in industrial markets in which it competes with other states. Thus, delegation of control to the states, however politically desirable, is frustrated by the need for such controls to be equitably imposed. This reality was heavily underscored in most states' failure to control coal strip mining, a situation which finally resulted in passage of the Federal Coal Surface Mining Control and Reclamation Act of 1977. Moreover, pollution, especially air pollution, is an interstate problem.

The federal government thus finally imposed federal air pollution standards in 1963 when the first Clean Air Act was passed. But imposed is too strong a word. This act was based on voluntary compliance and failed utterly. It took the Clean Air Act Amendments of 1970—federal, arbitrary, expensive, and probably not strict enough—to begin progress in cleaning up the air.

It should go without saying that the federal government itself is not without need for control. In fact, the National Environment Policy Act of 1969 (NEPA) was aimed specifically at controlling the abuses of federal highway and dam construction. While it received deserved criticism, namely that the requirement of an impact statement has not required that good substantive decisions be made, only that the procedure of decision making be specified and amenable to public participation, NEPA has made significant improvements in the planning of projects with potential environmental impact and has actually helped stop a few of the worst, such as the Cross Florida Barge Canal, an international airport in the Everglades, and possibly* the Tocks Island Dam on the Delaware River.

What does environmental law and/or the lack thereof portend for energy use? Should environmental laws be waived for energy projects? The answers lie in the tangled conjunctures of technologies, effluent standards, and health and property effects of energy procurement and conversion activities. It is important to understand that environmental control, while it constitutes a small cost within GNP (typically 2 percent) is a heavy cost to energy precisely because energy is a main locus of pollution problems.

6.3. Health and Property Damage from Energy Use

6.3.1. Air Pollution

The most troublesome air pollutants resulting from energy use are sulfur oxides (SO_x) and their derivatives such as acid sulfates, nitrogen oxides (NO_x) and their derivatives, carbon monoxide (CO), ozone (O_3), and

---

* This project has yet to be deauthorized by Congress.

particulates which include toxic elements like lead and unburned hydrocarbons. Sulfur oxides perhaps are the most troublesome not only because of their wide dispersal and destructiveness (when converted to sulfates and/or sulfuric acid), but also because sulfur is present in large quantities in coal, and in lesser amounts in oil. However, the automobile is the chief air pollutant offender as it produces most of the quantities of four of the above six "target pollutants" and does so in close proximity of people.

6.3.1.1. Sulfur Oxides

Electric utilities released from their smoke stacks about 20 million tons of sulfur dioxide in 1978. As sulfur oxides leave the stack and diffuse through the atmosphere, sulfate compounds are formed in a poorly understood reaction with particulates (aerosols) which escape electrostatic precipitation. The simplified reactions, which include the production of H_2SO_3 and H_2SO_4, sulfurous and sulfuric acid, are shown below:

$$SO_2 + (O) \longrightarrow SO_3$$
$$SO_3 + H_2O \longrightarrow H_2SO_4$$
$$H_2SO_4 + X^+ \longrightarrow XSO_4 + 2H^+$$
$$H_2O + SO_2 \longrightarrow H_2SO_3 \longrightarrow H^+ + HSO_3$$
$$\longrightarrow 2H^+ + SO_3$$

The X, above, represents metallic cations which may include vanadium, iron, manganese, lead, copper, and aluminum, which are components of the particulates.

These products may be transported over long distances. While New York has steadily reduced local sulfur emissions over the last two decades, the amount of SO_x, sulfates, and sulfuric acid present in its air has actually increased because of the influx of the pollutant carried by the winds from the nation's interior. Coal-fired power plant emissions in the Tennessee and Ohio Valleys are partly responsible for the fact that the pH of rain in New York and New England has recently been measured at below 3, and that gamefish have disappeared from 90 percent of New England's lakes.

Acid rain corrodes metals, concrete, limestone, marble, paint, diminishes the productivity of soil and natural vegetation, and attacks human lungs. Sulfates and acid damage the lungs by attacking the lung membrane and cilia in an incompletely understood way. It is thought that the motion of the cilia (hairlike cells which expel mucus, dirt, and germs from the lungs) may be slowed or even stopped by the effect of SO_2 and its derivatives as well as by the effect of other air pollutants. Thus, the lungs' resistance to disease is diminished. Air pollution irritants can also constrict the airways, thicken the mucus, paralyze bacteria-destroying cells in the respiratory system, and can induce swelling and excessive growth of the cells that form the lining of the airways. And because the heart and lung function together, the heart can be strained by trying to compensate for reduced air flow by pumping harder. Lung diseases can be accompanied by a doubling in the size of the heart.

The health effects of SO_x are devastating. Literally thousands of asthma attacks and cases of respiratory disease, and hundreds, perhaps thousands, of deaths per year are attributable to power plant SO_x emissions.[1]

6.3.1.2. Sulfur Oxide Standards

What are the standards? How good are they? The primary standard, set to protect public health, was to have been met by 1976. The FWPCA was amended again in 1977 allowing some deadlines to slip, but the primary standard of 80 micrograms per cubic meter (annual arithmetic mean) was retained. The standard for shorter periods of time (that is, maximum in a 24 hour period) is 365 micrograms per cubic meter. But the thresholds, the level of concentration above which adverse effects begin to appear are 91 and 300–400 micrograms per cubic meter, respectively (see Figure 6.1). Thus, a safety margin of only 14 percent is afforded the population for the annual standard, but no safety margin whatsoever exists for the short term standard. The effect of exceeding the short-term standard is what is morbidly called the mortality harvest. The harvest can be like that of Donora in 1948.*

The secondary standards, to be achieved by some as yet indeterminate date, would be a bit stricter at 60 and 260 micrograms of SO_2 per cubic meter over one day and one year, respectively. Note, though, that the standards apply to SO_2, not sulfates, and thus serve only as an index of safe sulfate levels.

The above standards apply to general air quality. Somehow, the administrator of the Environmental Protection Agency (EPA) must translate these into emission standards in order for them to be attained and equitably regulated. The EPA in turn delegates enforcement authority in state governments through State Implementation Plans (SIPs), but reserves the authority to decertify a state if that state's standards are not adequate. Meteorological modeling (and politics) plays a large role in determining

---

* The concept of "mortality harvest" implies that those who die in such cases as the Donora, 1948, inversion would have died soon anyway. Recent evidence indicates that victims are not necessarily such persons.

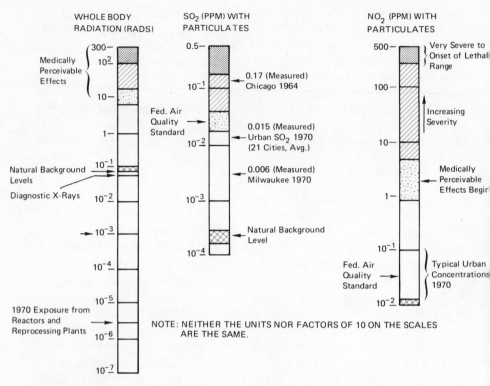

Figure 6.1. Comparison of pollutant standards, background levels, man-made exposures, and health effects. Source: Reference 9.

how much sulfur a given plant may release (or conversely, how much sulfur a plant's fuel may contain).

6.3.1.3. Sulfur Removal

Besides coal gasification and liquefaction, which we will consider separately, there are basically three approaches to coal cleaning. These are coal washing, flue gas desulfurization, and fluidized bed combustion.

6.3.1.4. Coal Washing

Coal washing usually involves crushing of the coal and its immersion in water. Materials such as pyrite, being more dense than the other constituents of coal, will sink in water, and the lighter coal can be removed off the top. About 40 percent of the sulfur content of eastern coals can be removed in this manner. Western coals have a lower ratio of pyrite to organic sulfur and thus washing is less successful. The cost of precombustion coal washing averages about ten cents per million BTU, or more than two dollars per ton. While washing eastern coal containing less than 1.2 percent sulfur by weight would produce "compliance" coal (under 1.2 pounds of SO_2 per million BTU), washing alone would not be the best available control technology.

6.3.1.5. Flue Gas Desulfurization (FGD)

In the U.S. in 1978 there were 30 operating flue gas desulfurization (FGD) systems and 86 under construction or planned. TVA alone has plans to order scrubbers for 3183 megawatts of plants to satisfy the Clean Air Act at ten TVA power plants. In Japan, 333 FGD systems have been installed, although most of these operate in small oil-fired plants.

Several criteria are used to evaluate FGD systems including cost, efficiency of sulfur removal, reliability, energy intensity, by-product production (that is, toxit and voluminous sludge vs. saleable elemental sulfur or sulfuric acid), and operating experience.

Lime, limestone, and double alkali systems are the cheapest in terms of operating and capital costs. The latter range from $75 to $125 per kilowatt of installed capacity. This cost may be placed in perspective by comparing it with an installed cost for coal-fired power plants—total, without scrubbers—of about $700 per kilowatt. But these systems remove only about 80 percent of the sulfur in the flue gas, and that which is removed must be disposed of in such a way as to not contaminate groundwater. A large fill area is required, for perhaps as much as 200 acres would be filled to a depth of 30 feet over the life of a 1000 megawatt plant. Operating costs run from $0.0025 to $0.005 per kilowatt hour compared with an average delivered residential rate of $0.035 per kilowatt hour.

Instead of a sludge as a by-product of scrubbing, the Wellman Lord system produces elemental sulfur which can be sold. This system currently has the drawback of having capital and operating costs twice as great as the nonregenerable lime and limestone systems. Significantly, natural gas is consumed in the elemental sulfur reduction process. A chief advantage is that the system can remove 95 percent of the SO_2. Other systems using magnesium oxide, aqueous carbonate, aqueous potassium, and carbon oxide may soon be available (MgO already is, and TVA will install several of these units, which produce sulfuric acid as a by-product) for a capital cost of $100 per kilowatt, with operating costs ranging from $0.005 to $0.009 per kilowatt hour. These systems typically remove 90 percent or more of the SO_2.

Many of the above systems, including the nonregenerable, sodium sulfite, and magnesium oxide systems have been demonstrated on a scale up to 100 megawatts and are reliable and effective. Much development remains to be done, especially in bringing down the costs of regenerable systems and finding ways to use coal rather than gas in the reduction of elemental sulfur. It should be pointed out, however, that a sizeable energy penalty is incurred in operating scrubbers, namely, a four to six percent reduction in saleable electric power.

6.3.1.6. Fluidized Bed Coal Combustion (FBC)

An amazing thing happens when air is blown at the right rate through the bottom of a container partially filled with some finely divided particles like sand. The solid bed material behaves like a liquid. It can be poured like a liquid and, most importantly, it has the high heat transfer characteristics of a liquid. Thus if after taking limestone and coal, crushing them both finely and placing them in a container through which air is blown upward to levitate the solids, one ignites the coal, several beneficial things happen. Very good heat transfer is facilitated to water pipes which may be immersed in the bed for the purpose of generating steam, and reaction of the limestone with the SO_x in coal prevents the SO_x from escaping into the atmosphere. In practice, most of the material in the levitated bed is inert ash. High mixing rates and high heat transfer facilitate very efficient SO_x removal (as calcium sulfate in the ashes) and high thermal efficiency. In fact, FBC boilers may need to be only one-fourth as large as conventional steam (or Rankine) cycle boilers per unit of output and thus can offer substantial savings in capital costs. It is as if economies of scale were reversed, because small units on the order of 10 to 100 megawatts and even smaller may be less expensive per unit of electrical output than very large conventional systems. FBC could then readily match cogeneration opportunities which tend to be in the tens of megawatt size range, especially if the hot gas of combustion can be used to turn a gas (Brayton) cycle turbine directly. The gas exhausted from the turbine can be channeled into a waste heat boiler for industrial process steam generation.

Other advantages include relatively low operating temperatures. A lower combustion temperature reduces NO_x formation and possibly suppresses the formation of particulates. The calcium sulfate by-product can either be used as a soil conditioner, or can be recycled to recover the lime and to produce SO_2 for sulfuric acid production. Sulfuric acid is worth about $25.00 per ton, although storage during times of low marketability is difficult.

An atmospheric-pressure FBC system (Rankine cycle) of 30 megawatts has been built by the Department of Energy in Rivesville, West Virginia, and is currently being tested. Meanwhile, a pressurized (Brayton and perhaps combined Brayton–Rankine cycle) FBC unit of 50 megawatts is being tested by Stal-Laval, Ltd., in England, for possible use in a 170 megawatt cogeneration system for the American Electric Power Company (AEP). Pressurized FBC systems could possibly be commercially available by 1985 or a few years thereafter. We return to FBC technology and its problems, particularly the potential for FBC application to cogeneration, in Chapter Ten.

6.3.1.7. Sulfur Oxides and the Future

Questions beg for answers. Where, for example, should coal come from in the future? How many people would become sick and/or die due to the effects of sulfur oxides if nuclear power were banned and replaced by coal? How much does SO_x control add to energy consumption and energy price?

Where should coal come from? Advocates of high energy futures would reply, "Everywhere, and as fast as possible." We do not believe this response is necessary. In Chapter Ten we present evidence that electrical energy demand growth will be slower than utilities have anticipated and argue that coal will not be needed in quantities greater than 1.5 billion tons per year until 2010.* (This ignores direct uses of coal, which are relatively small, and the use of coal for synthetic fuels, a subject discussed in Chapter Nine.) The eastern U.S. could meet this rate of 1.5 billion tons per year, we believe, and if western coal were needed, the alluvial valleys so important to ranchers and other westerners could be spared without tying up significant coal reserves. By retrofitting many old coal-fired steam plants with scrubbers (combined with coal washing for more economical use of the FGB systems), by washing and blending coals, by mining eastern low sulfur coals, and by relying on industrial cogeneration for new electric capacity, enough coal-derived power can be provided without making the arid Western coalfields "a national sacrifice area" as the National Academy of Sciences[11] has indicated they could become. (See water and land impacts of coal strip mining, below.) A shift toward longwall underground mining, a technology we discussed in Chapter Five, would help ameliorate some of the land and water problems which are inevitable with currently practiced methods of extraction.

What would be the health effects of increased exposure to SO_x due to a nuclear moratorium forcing increased reliance on coal? The effects would be nontrivial, but less serious than one might imagine, especially in a "low"

---

* The Department of Energy has recently revised its projections of 1985 coal demand downward from 1.5 billion tons to below one billion tons per year.

energy future. Suppose that no new nuclear plants were committed beyond 1980, that the U.S. consumed 100 quads in the year 2000 with half coming from electricity, and that emissions from coal combustion were limited to 1.2 pounds of SO_2 per million BTU. With these assumptions, about 14 million tons of sulfur would be emitted in the year 2000. Change only the last assumption to a requirement for scrubbers operating at 90 percent efficiency on all new plants, and the annual tonnage of sulfur emitted drops drastically to slightly more than 4.6 million tons. Under an alternative scenario in which no nuclear moratorium occurred but neither were scrubbers required, emissions might total 4.2 million tons, a total almost as great as the case of using the best controls with a moratorium. Dropping the moratorium assumption while maintaining the scrubber requirement reduces emissions to 1.4 million tons. For clean air, the single decision to require scrubbers appears to be more important than the moratorium decision.

How much energy is lost in SO_x removal? How much does SO_x removal add to energy cost? As stated previously, about four to six percent of the electrical output of a power plant is sacrificed to scrubbing, but this cost is added to the total cost which, even if the entire U.S. electrical capacity were retrofitted with scrubbers, would add no more than 17 percent to the total cost of electricity. This assumption, of course, overstates the cost drastically since only a fraction, the coal burning plants, would be retrofit. This cost would in fact be smaller than the cost differential of western and eastern low sulfur coal. That is, scrubbing would be cheaper than shipping western "low sulfur" coal to population centers because of the high transportation costs (at 1100 miles times $0.007 per ton mine equals $7.70 per ton, or approximately $0.50 per million BTU).[8] FBC, moreover, could be cheaper and more effective than either alternative.

6.3.1.8. Nitrogen Oxides

Whereas power plants are by far the largest contributor of SO_x, automobiles and power plants share the honor for NO_x. Nitrogen, as NO_2, can produce in human lungs much the same effect as SO_x. Nitrous acid mist is formed from NO in the presence of water. Formed at high temperatures in internal combustion engines and power plant furnaces, nitrogen oxides give the skies over polluted cities that muddy-brown haze that reduces visibility.

NO_x damages lungs by attacking the membrane and cilia and can impair the oxygen carrying capacity of the blood. Nitrogen dioxide can stunt vegetation growth, and HNO_3 (nitric acid), formed from NO_x in the presence of water, corrodes metals.

The Clean Air Act standard for NO_x (annual arithmetic mean) is 100 micrograms per cubic meter. The best estimate of a threshold for health effects (above which severe, acute respiratory illness increases) is 141 micrograms per cubic meter, a 41 percent safety margin. Most of the U.S. currently meets this standard, with the notable exceptions of Los Angeles, Denver, New York, Philadelphia, Chicago, Baltimore, Atlanta, Springfield (Massachusetts), Canton, Youngstown and Steubenville (Ohio), and Boston, all of which approach or exceed the limitation.

Reduction in combustion temperature was Detroit's first response to reduce NO_x emissions, and cars were engineered to operate at lower temperatures by means of recirculation of exhaust gas and reduced compression. Lower operating temperatures in internal combustion engines meant less efficient operation, however. The fuel penalty associated with these techniques came to about 10 percent, a large loss in the efficiency of an end use which consumes 15 percent of all American energy. The emission standards forced the reluctant auto industry to apply technology such as catalytic converters, which are installed in a third of all cars. Catalytic converters enable cars to meet clean air standards without significant fuel economy penalties. It should be cautioned, though, that presently marketed diesel engines, highly touted for efficiency, have a very high compression ratio and thus produce a great deal of NO_x as well as possibly carcinogenic particulates. The Clean Air Act Amendments of 1970 do allow a variance for nitrogen oxide emissions from diesel engines.

Some local governments have tried other air pollution control options, such as reduction of parking space (Boston), lower tolls for car pooling (Los Angeles), and blocking off streets entirely to automobiles. The object is to couple behavior modification with technical strategies to reduce simultaneously the emissions rate per mile and the number of vehicle miles.

As for power plants, reduction of combustion temperature is extremely difficult in conventional furnaces without severe fuel penalties and other systems problems. For new sources, as we pointed out previously, the fluidized bed combustor promises greatly reduced NO_x emissions because of lower operating temperatures and higher fuel efficiencies. If FBC were applied to cogeneration, emissions would be further reduced.

6.3.1.9. Hydrocarbons, Particulates, and Carbon Monoxide

Unburned hydrocarbons (HC) are problems for at least two reasons. They participate in the formation of NO_2, and they may cause cancer. The olefins, unsaturated compounds which react easily with other chemicals, are crucial actors in the oxidation of NO to NO_x because their presence is required for the formation of radicals which actually react with NO. Control of NO_2 formation is thus linked to control of olefins. Aromatic hydrocarbons such as benzpyrene are believed to be carcinogenic. The predominant source of olefins and aromatics in the atmosphere over the

U.S. is the incomplete combustion of fuel along the cold cylinder walls in the automobile engine. The installation of catalytic converters can effectively convert 90 percent of the unburned HC in exhaust to CO_2 and H_2O, and this efficiency can be further increased with relatively small marginal costs. Fuel efficiency costs, too, are small. The biggest problem is that the present generation of converters works poorly when cold and will not prevent emissions for one to two minutes after startup.[1]

Carbon monoxide in moderate concentrations can reduce the oxygen level in the blood and produce headaches, dizziness, fatigue, and confusion. Large amounts can cause brain damage and death. Although the overall ambient level of CO in the atmosphere is small, concentrations along heavily used streets can be serious. Catalytic converters can convert CO to CO_2 in the process of converting hydrocarbons almost without any extra expense or fuel loss.

Ozone is the most serious of the photochemical oxidants. In humans it can cause coughing, choking, eye irritations, headaches, and fatigue. Ozone stunts the growth of vegetation and deteriorates fabrics, rubber, paper, and other materials. It is a product and ongoing reactant in the formation of photochemical smog. Its control would seem dependent upon NO_x control.

Particulates may contain a variety of elements including toxic metals. A 1000 megawatt coal burning power plant may emit the following quantities of toxic metals per year (in tons): mercury, 5.5; beryllium, .44; arsenic, 5.5; cadmium, .0001; lead, .22; nickel, .55. Electric utilities point out that their electrostatic precipitators remove 99 percent of the particulates, by weight, in the flue gas. But the key is weight; over half the number of particles escape because they are too small to be precipitated. It is precisely these particulates (less than five microns in diameter) which can bypass nasal mucus and imbed themselves in the lungs. Further statutory emissions standards for particulates may be required to protect public health.

The existing statutory standards for HC, CO, O_3, and suspended particulates are shown in Table 6.1 with the "best estimate" of threshold levels and safety margins.

Note that the safety margin is very small for particulates and ozone. Particulate matter especially may prove to be a serious health problem.

6.3.1.10. Prevention of Significant Deterioration (PSD)

The Clean Air Act Amendments of 1970 settled upon uniform National Ambient Air Quality Standards (NAAQS). EPA's decision to enforce the NAAQS in a manner that allowed degradation of clean air areas aroused the ire of both environmentalists and judges who believed the

Table 6.1. Standards, Medical Thresholds, and Safety Margins for
Four Major Air Pollutants (In Micrograms per Cubic Meter, or Percent)

Pollutant	Primary standard	Medical threshold	Safety margin (%)
HC	160	Not available	Not available
CO	23(8 hr); 73(1 hr)	10(8 hr); 40(1 hr)	130(8 hr); 82% (1 hr)
O_3	200	160	−25
Particulates	250(yr); 100(daily)	260(yr); 75(daily)	None(yr); 33% (daily)
SO_2	91(yr); 300–400(daily)	(350 daily)	None

Source: Reference 1.

purpose of the Act was to "protect and enhance" [§101(6)(1)] air quality. Thus the prevention of significant deterioration (PSD) regulations were born.

Specifically endorsed by Congress in 1977, the PSD regulations divide the airsheds of the country into three classes. Class I areas are pristine areas like national parks, wilderness areas, etc., where no increased pollution will be allowed. Class II areas will be allowed such increments of increased pollution as ordinarily accompany moderate growth. Class III areas will be allowed the maximum amount of growth that will not push pollution levels beyond the NAAQS. Most of the country, except the pristine areas which were designated mandatory Class I, was designated Class II, and it will be up to individual states, with extensive help from EPA, to redesignate areas as Class I or Class III, depending upon the type of development they desire.

These standards will mean better health and fewer headaches for the population, but more headaches for industry and EPA. It has yet to be determined how the remaining permission to pollute will be parceled out to new or expanding industries including electric utilities. It is clear, however, that areas which fail to regulate automobile emissions may find that industrial expansion is choked off.

6.3.1.11. Synthetic Fuels and Air Pollution

Coal and oil shale conversion deserve special consideration relative to air pollution. As Table 6.2 indicates, all of the aforementioned air quality factors plus some others make coal gasification and/or liquefaction serious environmental business. The table indicates considerable lack of knowledge about the health effects of these potential emissions. It should be pointed out that oil shale processing has a comparable potential for air pollution disaster. To the extent that conservation can reduce energy demand, the need to switch to synfuels with all their possible deleterious side-effects can

Table 6.2. Potential Health Impacts of Emissions from Coal Gasification and Liquefaction Plants

Higher significance	Lower significance
More poorly understood	

Benzene: a known carcinogen

Beryllium: suspected to cause bone and lung cancer

Cadmium: possible relation to prostate cancer

Fluorides: may increase sensitivity to chemicals affecting central nervous system

Lead: suspected occupational carcinogen

Nickel: occupational cancer incidence

Nickel carbonyl: causes lung cancer, possibly asthma

Nitric acid: can irritate eyes, lungs, mucous membranes, skin, and corrode teeth

Nitric oxide: can cause pneumonia, circulatory system damage; suspected respiratory irritation and tooth corrosion

Nitrogen dioxide: suspected to reduce resistance to bacteria; acute exposure causes increased respiratory inhibition

Phenols and cresols: occupational carcinogen (skin): may damage central nervous system and liver

Selenium: occupational cause of digestive and nervous disorders

Sulfur dioxide: correlates with chronic respiratory diseases; synergistic effects with particulates

Zinc chloride: possible carcinogen

Carbon monoxide: suspected to alter enzyme activity; cause behavioral changes; precipitate heart attacks

Fluoride: suspected association with blood disorders

Manganese: causes brain damage and pheumonia in high doses

Xylene: inhibition of electrical activity in cerebral cortex at levels below odor threshold

Vanadium: acute respiratory irritation; chronic ingestion produces systemic symptoms

Zinc oxide: occupational exposure can cause intestinal, respiratory, skin, and nervous disorders

Better understood

Beryllium: causes acute and chronic respiratory disorder from short-
   term exposure
Chromium: suspected cause of lung cancer
Fluorides: high levels lead to chronic poisoning or fatality; can cause
   respiratory impairment
Lead: damages central nervous system
Polycyclic aromatic hydrocarbons: carcinogenic
Uranium: insoluble compounds damage lungs; salts damage kidneys
   and arteries

Arsenic: lethal at high doses
Barium: eye, nose, throat, skin irritant; salts and sulfide poisonous
Beryllium: causes chronic berylliosis
Cadmium: systemic and fatal effects from inhalation
   of high concentrations
Carbon monoxide: causes dizziness, fatigue, and coronary
   dysfunction
Chromium: occupational exposure causes lesions of skin and
   mucous membranes
Cyanides: high concentrations lethal
Phenols and cresols: corrodes skin and mucous membranes
Selenium: causes dermatitis and respiratory irritation
Toluene: chronic exposure can cause brain damage

Source: Reference 3.

be delayed or reduced. We believe we must not waive environmental laws designed to protect people from sickness and death resulting from energy production. To do so would be pernicious.

6.3.1.12. A Word on Thresholds

The concept of thresholds of health effects embodies a continuing debate. On one hand, scientists argue that no threshold can be determined because statistically it is impossible to assess the effects of pollutants at very low levels. This argument, if valid, would destroy the logic which helped establish statutory ambient standards. On the other hand, the human body has repair mechanisms which enable it to tolerate exposure to many dangerous substances below certain levels. The "no-threshold" argument usually leads to the assumption of a linear extrapolation of observed effects at measurable doses down to levels for which observations cannot be made simply because not enough animals can be experimented on to generate valid statistics. Many argue that statutory standards are irrelevant and that they should be replaced by benefit/cost calculations to determine the most cost-effective effluent limitations.

But the problems with cost/benefit analyses are great. As the history of National Environmental Policy Act (NEPA) illustrates, cost/benefit calculations are no better than the assumptions which go into them, and worse, these assumptions are subject to manipulation. While economic theory requires that total costs and benefits be included in cost/benefit calculations, in practice these calculations have little utility when the benefits to be quantitied include human lives and intangible values. To be sure, a ranking of those environmental goals we wish to obtain is in order. We cannot afford zero pollution. Still, the proper forum for such determinations, we believe, is not the cost/benefit calculation *per se,* but representative bodies such as the U.S. Congress. The progress we have made toward cleaner air, and we have made some, indicates that enforceability derived on the basis of benefit/cost analysis, is the key to environmental standards.

6.3.1.13. Air Pollution and Energy: A Summary

The effects that increased electrical generation and automobile use will have on air quality and human health can be profound. Existing statutory standards represent an effective if not perfect first step toward cleaning up. The economic costs appear to be sustainable and the benefits will be great. The constraints on energy development will not be disastrous, though restrictions on industrial and energy supply development could be reduced if offset by stronger controls on automobile emissions. Particularly effective would be an increase in automobile fuel economy. Further investigation into the health effects of toxic metals and other effluents is urgently needed. Fortunately, most energy conservation options will reduce the health and property damages of air pollution.

6.4. Carbon Dioxide: A Global Threat

The atmosphere of the Earth, barely twelve miles deep, is unique, and it is an understatement to say it is not well understood. Earth is the only known planet where water can be liquid a substantial portion of the time. Life as we know it depends on the substantial quantities of liquid water made possible by Earth's unique atmosphere. On Venus, carbon dioxide exists in the atmosphere in such quantities that reduced amounts of the sun's radiation are reradiated to space. The CO_2 blocks infrared (heat) radiation like the glass in a car, although it is transparent to shorter (light) waves. The resulting heat buildup makes life as we conceive it impossible.

The question of whether the well-established buildup of CO_2* in the Earth's atmosphere will ultimately result in climate and weather modification, ice cap melting, or worse catastrophies cannot be answered yet because of several major uncertainties. First, we do not completely understand how the ocean removes CO_2 from the atmosphere and stores it, and there is some possibility that the ocean's capacity to serve as a CO_2 sink is being saturated. Second, we cannot accurately project the future of forests and soils which also effectively store CO_2. Also, we have shown in the first part of this book that we cannot predict with confidence future rates of energy use. Finally, we understand too little about the atmosphere itself to predict its behavior. We do know, however, that the old argument that particulates thrown into the atmosphere could effectively counteract a heat buildup by reflecting the sun's light is a poor one, because such a quantity of particulates would probably so drastically change the climate and weather patterns that disaster would result.[9]

Uncertainty about CO_2 buildup is highlighted by the fact that only half of the CO_2 produced since 1958 has actually accumulated in the atmosphere. Whether the natural "sinks" for CO_2 (such as the ocean) will become saturated and drive up the rate of accumulation cannot now be decided. Rapid clearing of tropical rain forests, for example, might con-

---

*For a discussion of this dilemma, see Stephen Schneider, *The Genesis Strategy, Climate and Global Survival,* Plenum Press, New York, N.Y., 1976.

tribute significantly to this accumulation. Although mature forests release as much carbon dioxide as they consume, an enormous quantity of carbon dioxide is stored in the cellulose of which these forests are composed.

In terms of CO_2 produced per unit of energy released from the combustion of fossil fuels, coal is the worst offender and natural gas the least. A scientific consensus is emerging that while we do not know what might happen as a result of CO_2 buildup, important worldwide effects could commence as early as 30 years from now, and 30 years is already less time than is needed to do something about the problem. The U.S., consuming fully one-third of the Earth's commercial energy, could help delay this crisis by conserving CO_2 producing fuels.

6.5. Radioactivity: A Special Pollutant

The health effects of radionuclides stem chiefly from their capacity to damage human tissue. For instance, in the relatively rare event of a chromosome being struck, information carrying chemicals in the chromosome may reform in deleterious ways.

In the normal operation of a nuclear power (or fuel reprocessing) plant, two isotopes which are particularly difficult to filter, tritium and krypton-85, are troublesome. Tritium, with a half-life of 12.3 years, survives long enough to exchange with hydrogen (with which it is chemically identical) in water, which may thus be ingested by animals. Since tritium is a beta emitter of relatively low energy, and since the time that the body retains water is relatively low (short biological half-life), tritium is less bothersome than it might otherwise be. Radioactive krypton-85, despite its danger to Superman, is chemically inert, but may deliver radiation to the lungs when inhaled since it will reside in the atmosphere as a noble gas. The question, to recall the threshold argument, is, "How much (or little) is safe?" Traditionally, the question has been answered in terms of comparison with natural background radiation which equals about 100 millirem per year. In contrast, a dental X-ray exposes one to about 20 millirem, while an abdominal procedure (4 X-rays) can equal a 5,000 millirem exposure. Radiation sickness is caused by a whole body dose of 100,000 millirem in a period of one month, and fifty percent of the people exposed to 500,000 millirem over their whole bodies in a single dose will die within 30 days, the lethal dose (LD) being 50/30.

The standard for incremental public exposure, which includes aquatic and atmospheric sources, has been set far below natural background in part because the costs of attaining such low release levels are also low. The standard is approximately one millirem per year from nuclear power, or 170 millirem from all sources.

The main concern for nuclear safety, aside from the proliferation of nuclear weapons, is the possible event of a catastrophic release from an operating plant or waste facility. Two recent accidents or near catastrophes illustrate that such concerns are well founded. In 1975 at the Brown's Ferry* nuclear plant, a fire induced by a workman's candle flame eliminated all remote control of the plant. A period of two days passed before the fire was extinguished and control of the plant regained. At Three Mile Island, Pennsylvania, an incredible series of events led to a partial core melt. If the accident at Brown's Ferry was not reassuring, the Three Mile Island fiasco was surely a near catastrophe. At the very least, the financial loss at Three Mile Island was so severe that investors now regard nuclear plants as very dubious investments. For that reason as much as any other, the nuclear option may be lost.

The magnitude of a major release such as that which occurred at a nuclear weapons waste facility in Russia in 1956 or 1957 has not occurred in the U.S. The incident in Russia was a chemical explosion which caused the dispersal of radionuclides over a 50 by 20 mile area. While a chemical explosion of that nature is unlikely in nuclear power systems because such compounds are not produced, a core melt could release enough steam and thermal energy to propel heavy radionuclides over such distances. The consequences of such a release, of course, would depend on the density of the population exposed.

The issues of long-term storage of nuclear wastes, their transportation and reprocessing, remain of great concern. Permanent isolation of these wastes will require their storage in dry, geologically stable underground formations. Salt deposits have been proposed for storage of nuclear wastes because salt deposits have a long history of stability and they are relatively plastic. Their plasticity would tend to keep the wastes in place if the salt layer moved with geological stress. Giving absolute assurance that the salt will be dry, however, and that the waste will not be transported to the surface over a period of thousands of years, is impossible.

6.6. Mining Atmospheres and Worker Exposures: An Occupational Air Issue

Pneumoconiosis costs the U.S. taxpayer one billion dollars per year. How this translates into suffering for those coal miners who develop black

* A Tennessee Valley Authority nuclear power plant in Alabama.

lung is incalculable. Essentially, the membrane of the lungs is destroyed by the deposition of coal dust during mining operations. The rate of disabling of miners because of black lung has been estimated to be between .5 and 7 miners per year per 1000 megawatt coal burning electric-generating plant. The total number of persons who would be disabled annually in the mining of 1.5 billion tons of coal per year could total between 150 and 2000 even if it is assumed that half of the coal is surface mined. Technology for virtually eliminating miners' exposure to dust has existed for years, but no effective effort to require the use of such equipment is apparent from government, industry, or labor. Black lung can be serious in surface mines too. Disabling cases presently number .5 per 1000 megawatt plant per year.

Uranium miners, though there are far fewer of them per unit of energy produced, suffer a similar personal risk. Radon is an inert gas which, as a decay-product of uranium, accumulates in uranium mines. It is an alpha emitter. Inhaled, alpha emitters can damage the lungs. The risk to a miner (who smokes cigarettes)* of developing cancer is almost ten times greater than for the average population. This translates into a rate of .1 disabling cancers per 1000 megawatt reactor per year.

6.7. Water

Energy development has affected water seriously. Hundreds of once free flowing rivers are now silted reservoirs behind hydroelectric dams. Ten thousand miles of streams in Appalachia run yellow, red, or black with silt, acid, or coal solids washed out of strip mines and coal preparation plants. And in the heyday of the FPC-style projections of electric power growth, it was forecast that by 1990 the equivalent of two-thirds of the daily runoff of water on the contiguous 48 states would be required for cooling power plants, and that this water would be heated by 10 °F over ambient.

6.7.1. Hydroelectric Impacts

Major impacts of large hydroelectric projects have been social and include the forced removal of families who have held land for generations and the destruction of their land by innundation. The areas lost to such projects often have been prime farmlands and productive forests, and the continued loss of such land should be of concern to a nation that is increasingly dependent on agriculture for both food and major exports.

---

* For a nonsmoking miner, the risk is *five times* lower: "only" twice that of the average population.

The famous snail darter case illustrates well the destructive side of hydroelectric power. Fifteen thousand acres of farmland were removed from production, families were forced from their homes, and a species lost its only known habitat.* For millenia the snail darter lived in the waters of the Tennessee River system, but its habitat had been reduced systematically until the fish was confined to a mere 33 miles of river. The Tellico Dam has now eradicated the last miles of the darter's habitat. The dam itself is small and has no generators; however, by diverting water through a canal to another nearby reservoir which itself is dammed by an installation having excess capacity, 22 megawatts of power-generating capacity has now been added to the TVA system. But it is a system that already has 30,000 megawatts.† The impact of such reservoirs on freshwater aquatic life has been devastating, especially for shellfish. It follows, then, that any effort to add a few kilowatts of capacity to a system by the construction of small-scale hydrodams must be scrutinized very carefully. Such scrutiny was part of the intent of Congress in passing the Endangered Species Act of 1973. This Act was meant to serve as an indicator of valuable habitats, as well as to prevent the rapid extinction of species, which is occurring at the rate of 5000 per decade. Enforcement of the Endangered Species Act usually requires modification of a project rather than its abandonment. Grayrocks Dam provides a good example. Completion of the project was allowed as long as assurance was given that enough downstream water flow would be maintained to sustain the nesting grounds of the endangered whooping crane. Proper consultation with the Federal Office of Endangered Species as mandated by the Act will protect endangered and threatened species without preventing the proliferation of energy-related projects.‡ Yet, the Endangered Species Act is vulnerable to ill-conceived efforts to waive Federal Laws to rush completion of energy projects. A law was passed to exempt Tellico Dam from all Federal laws in order to complete it. It was argued that TVA needed the energy. The irony of this argument is that TVA concurrently postponed completion of 5000 megawatts of power plant capacity for five years because TVA did not need the power. Yet all Federal laws were waived to complete 22 megawatts. Tellico Dam is the epitomy of a new form of energy demagoguery that must be faced and discredited.

---

*The snail darter *(Percina tanasi istoma)* requires cool, clear flowing water to survive and gravel shoals to reproduce. These qualities have been eliminated by impoundment.

†TVA adds electric-generating capacity at a rate equivalent to installing a Tellico every week on the average.

‡More than 5000 potential conflicts with the Endangered Species Act have been resolved administratively. Only three have gone to the courts, and only one, the *Snail Darter* case, has had to be resolved by Congress.

6.7.2. Mining Effluents

The impact of mining on water is severe. It includes erosion of silt from disturbed land, sulfuric acid drainage, leaching of toxic metals, flooding when streams fill with debris, and the destruction of aquifers and wells. The Surface Mining Act of 1977 was passed to correct some of the worst abuses of coal surface mining and to establish a fund to perform minimal reclamation to stabilize the two million acres of abandoned coal strip-mined lands which continue to cause water pollution.

Controlling erosion from strip mines is difficult. Sedimentation ponds help, but still erosion may be 1000 times as great from a surface mine as from a forest floor. Carried out of a strip mine along with the sediments are toxic elements such as cadmium, zinc, arsenic, copper, lead, and mercury, all of which may attack the respiratory systems of aquatic life and/or end up in human water supplies. Sediment basins allow water to collect and permit some of these effluents to precipitate out before the water is discharged to a stream. Acid, formed by the oxidation of pyrites by the action of air, water, and bacteria, will effectively sterilize a stream unless discharge water is diluted sufficiently. Naturally, retaining large amounts of water in sediment ponds on steep Appalachian slopes in the humid east is difficult and dangerous. While the statutory requirement of returning land to its approximate original contour (and burying of toxic material) as contemporaneously with mining as possible will reduce erosion and leaching, continued serious water pollution can be expected from mining slopes greater than 15° and from mining operations that remove mountain tops. In light of the fact that only very small amounts of coal are mineable under such conditions, it would seem prudent to phase out rapidly such destructive methods.

In the arid west, the problem is too little water rather than too much. Since many of the coal seams contain aquifers from which ranchers obtain domestic and agricultural water, Congress opted to prohibit mining in alluvial valley floors where the aquifers run. The language of the law is so vague, however, that this resource may not be fully protected. Again, aside from the fact that the need for western coal may be obviated, the coal resource which lies in these aquifers is relatively small, making protection of the water a small loss to U.S. energy supply.

In deep mines, acid drainage and coal washing discharge can be serious water pollution problems. Recently, the Federal Office of Surface Mining (which has jurisdiction over all coal mining water discharges) has moved to force the industry to apply simple control techniques to prevent such discharge.

The cost of the federal strip mine controls has been estimated by several sources. In the west, where the number of tons of coal per acre is large but the BTU content per ton is low, the additional cost of the new federal reclamation requirements should run about $0.02 per billion BTU, compared with $0.05 in the midwest, and $0.11 per million BTU in Appalachia.[6,8,9,11,12] In Appalachia, this cost should be added to that of about $0.10 per million BTU spent meeting existing requirements. Added to this is the tax assessed for the reclamation of abandoned lands, equivalent to about $0.02 per million BTU. These costs compare with the 1979 prices of coal delivered to electric utilities of almost $1.00 per million BTU and delivered costs of electricity from coal-fired plants of $10.00 per million BTU. Overall coal production losses* due to the Act have been estimated to be less than two percent.[12] The total cost of the strip-mining Act is less than one percent of the cost of electricity.

The production of oil shale would not only require tremendous quantities of process water, but would carry the threat of polluting surface and ground water with toxic effluents from disposed spoil. The disposal of one to two tons of fractured shale per barrel of oil produced will present a tremendous surface area for the erosion of such materials.

6.7.3. Gasification and Liquefaction

Many of the airborne emissions listed in Table 6.2 as potential air pollutants from gasification and liquefaction are potential water pollutants, especially the inorganic compounds. Many of these are powerful carcinogens. Others are toxic.[13] In addition, direct water consumption would total three to five barrels of water per barrel of oil produced.

6.7.4. Oil

The impact of oil on living systems can include:

o   death as a result of coating and asphyxiation
o   death through contact poisoning
o   death from exposure to water-soluble toxic compounds
o   destruction of sensitive juvenile forms
o   destruction of food sources of support populations
o   incorporation of sublethal amounts, resulting in the reduced resistance of species to infection or stresses
o   incorporation of sublethal amounts, producing an off-flavor or taint in exploitable species, thereby causing an economic loss to man.[14]

---

* These losses reflect the percentage of marginal mine operations that would be forced out of business.

Oil tanker accidents represent only the tip of the iceberg, so to speak, of oil pollution. Automobile waste oil accounts for 26 percent, industrial operations and routine tanker discharges account for 22 percent, natural seepage for 15 percent, vessel accidents for 5 percent, and offshore oil drilling for 2 percent of the oil discharged to American waters. Since much of this approximately *28 million barrels each year* is nonpoint source discharge, it will probably not be subject to control in the foreseeable future.[2,14]

6.7.5. Thermal Pollution

Many new power plants will have to recycle their cooling water rather than return it to its source in a once-through cycle. EPA estimates, however, that perhaps 70 percent of new steam-electric plants could avoid the requirement of a closed-cooling cycle. EPA regulations (pursuant to the Clean Water Act) require that no more than two percent of a power plant's heat be discharged into water, that no more than 100 cubic feet per second be withdrawn, and that no species of aquatic life be harmed. But these criteria may be meaningless since so many site-specific variables like water flow and ambient temperature determine the overall effect.[1,2]

Of concern are the decrease of dissolved oxygen in water as temperature rises, the increased adverse effect of chemical pollutants at elevated temperatures, and the physical sensitivity and behavioral changes in aquatic life in response to temperature change. Probably the most infamous episode occurred when the Oyster Creek (New Jersey) reactor shut down in winter and caused 500,000 Menhaden to die. Normally, the fish would have migrated south to warmer water, but they had failed to note the change in seasons because of the warm discharge from the power plant. The sudden, cold shock that followed the plant shutdown killed them.

The cost of cooling towers to reduce direct thermal discharge into water bodies might run as high as a few billion dollars nationwide by the mid-1980s. An energy penalty on the order of two to six percent results as well. It would seem that site- and season-specific flexibility would be prudent, and is embodied in the law, in making decisions about closed-cycle cooling. Of more importance perhaps is the control of discharge of chemicals such as chlorine used to treat the cooling water. Also, if very high growth rates in power capacity are realized, then closed-cycle cooling could become essential in some areas.

6.7.6. Water and Energy, Summarized

Strip-mining effluents and aquifer disruption remain a serious problem despite attempts at federal control. Oil pollution of both inland and ocean waters, already considerable, is growing. The effects are serious, but control efforts are weak, especially on the numerous sources of our fresh and estuarine waters. Thermal pollution may become a serious problem if high electric power growth rates are realized. Otherwise, the impacts of heating natural water bodies may or may not be serious, depending upon the site. Generally, water availability will not only limit coal production and conversion in the west, but could also cause serious local impacts in the east.

Each of these problems is serious; our attention to each should be maintained.

6.8. Land

As should already be clear, many of the issues of energy and environment cut across air, water, and land categories. Strip mining and power plant siting are obvious examples, and in choosing how we use our land, two additional issues arise, power lines and urban sprawl.

6.8.1. Mining

We have previously cited the costs of reclamation which are relatively small compared to the price of coal. Reclamation costs, however, may exceed $5000 per acre in areas where land sells for only $200 per acre, and so one is tempted to ask, "Why do it?"

Part of the answer is that external costs to local and area residents need to be prevented. The damage inherent in such costs is, in our opinion, a national disgrace. And it was not aesthetic concerns that generated the first efforts at control. Remember that in the recent past spoil was shoved over the side of the mountains. The damage to land and buildings from slides and floods caused by creeks filling with sediment and spoil generated the first efforts at control. Blasting cracked foundations and ruined wells. Outmigration from communities near strip mines was and remains high. The number of jobs generated by surface mining is low.

Spoil left unburied may develop a pH of three. Toxic metals leach out because of the acid and contribute to the prevention of revegetation. And even if topsoil is returned, the land may not recover its prior productivity because stored topsoil leaches and is compacted. Microorganisms, so vital in fertile soil, may be destroyed. In the arid west, reclamation is far more difficult; it is virtually impossible in areas that receive less than ten inches of rainfall per year.

In short, reclamation is tenuous. In fact it is a misnomer, for rehabilitation, especially on steep slopes, is the best one can expect. Fortunately, it may not be necessary to continue to mine coal on steep slopes, or to shift to the west for coal supplies.

Land is presently being disturbed by strip mining at a rate of 100 square miles (66,000 acres) annually. To the extent that coal consumption and/or strip mining production grows, this number will grow. Underground longwall mining is not without surface effects since subsidence can sometimes be severe. But the sooner we get on with the inevitable job of shifting coal production back underground (recall that as much as 90 percent of the U.S. coal resource may be recoverable only by deep mining), the safer and less destructive it will be.

6.8.2. Powerlines

One of the chief objections to electric power growth is the need for those leviathan transmission towers and broad rights-of-way. Objections to the potential health effects of high voltage transmission have recently become strident, particularly in rural Minnesota. Examining scenarios of future power line requirements, we wonder if this land-use issue has not received far too little attention.

If electricity does supply one-half of all U.S. energy needs in A.D. 2000, and that level of (total) demand equals about 100 quads, then 16.6 million acres, an area almost equal to the present national park system, will be required for electrical transmission.[9] It is by no means certain, however, that such a huge proportion of our total energy supply will be electrical. Neither is it true that all lands associated with transmission rights-of-way are taken out of use. Nevertheless, the aesthetic prospect is overwhelmingly bad.

6.8.3. Urban Sprawl

Urban sprawl is tangentially an energy issue. Certainly the segregation and wide separation of residential, commercial, and work areas contributes mightily to our oil import and clean air problems by requiring so much travel. And if industry is constrained by the unavailability of pollution rights, the lack of margin between existing ambient air quality and air quality standards, then the automobile which both made sprawl possible and is made essential because of it will be in very large measure to blame. Electric autos may or may not help on this score; they may simply transfer the emission of pollution to a single point.

While only peripherally an energy issue, sprawl may be even more fundamentally related to the environmental revolution. Sprawl is ugly.

Buildings which "look like they weren't made to last very long,"* the sterile asphalt seas, the signs, the tangle of wires—these, perhaps as much as smog and water pollution, cause human discontent.

6.9. Summary

We have highlighted some of the external costs of energy production and utilization. Energy use, more than almost every other human activity, affects environmental quality. One major point we have tried to make is that to ignore environmental and health costs in order to have cheaper energy is to obtain a false bargain. Insofar as the cost of saving energy equals or even exceeds the supply "price," it can still be cost-effective. The following chapter develops this idea more fully.

* See *In Dubious Battle,* by John Steinbeck.

. . . in the long term, to protect the consumer from the reality of the energy crisis is to destroy him.

—SENATOR PAUL TSONGAS
Dem., Massachusetts (1980)

Chapter Seven

Energy Supply Policy

7.1. Confluence

The metaphor of a watershed aptly describes the energy situation. Disparate opinions and branches of information about energy are beginning to flow together toward a consensus. Gramsci's image* of the hiatus between the lingering death of an old system and the difficult birth of the new also rings true. The new emphasis on energy conservation is almost literally a revolution. But a watershed is most graphic. With the prevailing divergence of opinion and disjuncture of information, it is difficult to organize a body of knowledge and thought into a single channel of ideas. We hope that in this chapter we can bring together information on the economic, environmental, and physical issues of energy supply in order that better policies, a better consensus, can be formed.

*See page 235.

A brief recapitulation of certain energy supply schemes, a review of thermodynamics as it relates to energy policy and the interchangeability of energy carriers, and an accounting of certain external energy supply costs as well as the economic costs will help organize this information for acting on the issues.

7.2. Projected Energy Supply

The energy demand scenarios reviewed in Part I had their supply as well as demand scenarios, and we briefly compare them now. The differences are great, both in comparison with each other and in contrast to energy supply today.

7.2.1. A Perspective on Current Energy Supply

Out of about 80 quads of annual energy use in the U.S., 40 quads are supplied by oil and 20 by natural gas. Twelve quads of oil are imported. Coal supplies about 15 quads, with 80 percent going for electric generation. Nuclear electricity and hydroelectricity supply one to two quads, and biomass (wood waste) adds one quad (see Table 7.1).

Translated into physical units, annual consumption totals:

o   more than six billion barrels of oil
o   20 trillion cubic feet of natural gas
o   775 million tons of coal

The capital equipment for supplying U.S. energy is:

o   60 gigawatts (60,000 megawatts) of nuclear power; that is, the equivalent of sixty large power plants
o   210 gigawatts of coal-fired power plants
o   200 gigawatts of oil-fired power plants
o   70 gigawatts of hydroelectric dams
o   550 gigawatts of power plants, total
o   No synthetic fuels plants
o   517,000 oil wells
o   157,000 natural gas wells
o   6500 coal mines
o   287 oil refineries
o   175,000 miles of petroleum pipelines
o   over 100,000 miles of electric transmission lines as well as thousands of coal trucks, gasoline tankers, rail cars, support facilities, etc.

Table 7.1. Energy Supply Components

Energy source	1978 actual	USBM-1975 for year 2000	iea-1978 for year 2000	CONAES-1979 for year 2010 "A"	CONAES-1979 for year 2010 "B"
Domestic oil: quads (billions of barrels)	20 (3.4)	27 (4.7)	21 (3.6)	28 (4.8)	28 (4.8)
Imported oil: quads (billions of barrels)	20 (3.4)	24 (4.1)	7 (1.2)		
Natural gas: quads (trillions of cubic feet)	19 (20)	20 (20)	18 (18)	9 (9)	11 (11)
Coal: quads (billions of tons)	15 (.7)	35 (1.8)	31 (1.6)	NA	NA
Solar: quads	2	6	3	NA	NA
Nuclear: quads (thousands of tons of uranium)	1 (8)	46 (130)	15 (43)	NA	NA
Oil shale: quads (billions of tons)	—(—)	6 (1.25)	7.5 (1.6)	NA	NA
Total: quads	77	163.5	101.5	74	94
Conversion facilities					
Nuclear power plants: number of plants (gigawatts)	70 (60)	900 (900)	300 (300)	NA	NA
Coal-fired plants: number (gigawatts)	—(211)	500 (500)	570 (570)	NA	NA
Electrical plants: total number (gigawatts)	—(552)	1900 (1900)	1000 (1000)	—(380)	—(490)
Coal synfuels plants: number (million barrels per day)	—(—)	90 (1.4)	—(—)	NA	NA
Oil shale synfuels plants: number (million barrels per day)	—(—)	20 (1.5)	60 (3.1)	NA	NA

Scenario[a]

[a] NA = not available.

A refined barrel of oil yields about 20 gallons of gasoline, 12 gallons of diesel, kerosene, and heating oil, and eight gallons of heavy fuel oil, asphalt, etc.

A cubic foot of natural gas contains 1000 BTU, approximately, or energy enough to heat water for one hot shower.

A ton of coal contains about 22 million BTU if it is good, eastern coal, and about 14 million BTU if it comes from the west. Since a barrel of oil contains about 5.8 million BTU before refining, a ton of eastern coal contains 3.8 times as many BTU as a barrel of oil.

An average large power plant generates a gigawatt (1000 megawatts) of power in any instant, or about .02 quads of electricity per year (.06 quads of heat), and serves about half a million people.

A plant producing synfuel from coal would produce between 30,000 and 50,000 barrels of oil per day. About one ton of coal would be required for each barrel of oil if the process were no more efficient than the Fischer–Tropsch process which is only 30 percent efficient. Thus, a million barrels of oil per day from coal would require 365 million tons of coal per year, an amount of coal almost equal to half our current annual production.

Oil shale plants might be three times as large as coal synfuels plants and would require perhaps two tons of shale per barrel of oil. Thus, a million barrels of oil per day from oil shale would require the mining of 730 million tons of shale each year, a quantity almost as great as current U.S. coal production.

With these perspectives in mind, we now turn to projections of future energy supply.[1]

7.2.2. Energy Supply in the Year 2000

7.2.2.1. As Projected by the USBM-1975

The USBM-1975 scenario, still widely cited, will serve as a typical "business as usual" scenario. The demand level of USBM projected totaled more than 160 quads in the year 2000, double current demand.

To meet this demand level, the USBM-1975 authors predicted that oil consumption would increase to 50 quads or more, and that half of this oil would be imported. An amount of natural gas equal to that produced today would be required. Coal use would increase by 2.3 times the current level to almost 1.6 billion tons per year. Nine hundred large nuclear plants and 300 large coal-fired power plants would have to be built and operated. In all, 1900 gigawatts of power plants, or 1900 large generating stations, would be in operation in the year 2000 compared to 550 gigawatts today. Plus, 87 coal conversion plants and 30 oil shale retorts would be required.[2]

7.2.2.2. As Projected by the iea-1978

The iea-1978 projection serves as a good scenario of relatively low energy growth with emphasis on a transition to an electric energy economy. Total demand, as we reported in Part I, was estimated to be about 100 quads per year in the year 2000. Since the iea study's purpose was to examine the impacts of a nuclear moratorium, the authors were assuming that no new nuclear plant construction would begin after 1985. So, of the 1000 gigawatts of power plants envisioned by the year 2000, 300 would be nuclear, though 1000 gigawatts would be sufficient to generate 15 quads of electricity. In the iea nuclear moratorium scenario, the number of coal-fired plants was more than doubled, as was the total use of coal. No coal-synfuels plants were projected, however, because of their high cost. Oil from shale was predicted to make a major contribution of 6.5 quads in the year 2000. Thirty plants and more than a billion tons of shale would be required. Solar energy used directly would provide two quads.

Interestingly, iea also forecast a level of natural gas production at about that of today. Oil use, however, would decline by a third.[3]

7.2.2.3. As Projected by CONAES-1979

The results of the CONAES-1979 demand study indicate much less dependence on electricity than did the iea project. Total year 2010 electrical energy supply was projected to be about 25 to 32 quads, or 380 to 490 gigawatts for the lower demand scenarios "A" and "B" (74 and 94 quads of total energy demand, respectively). This indicates diminished demand for electrical capacity.* Total electrical output would be 5 to 12 quads greater than today, a discrepancy accounted for by far higher capacity utilization. Also according to scenarios "A" and "B," liquid fuels use ranged from about the same as today to a 30 percent decline. Natural gas use would decrease by up to 50 percent, however. Solar energy, was not differentiated from energy conservation in terms of supply.

Table 7.1 summarizes and compares the above information.[4]

7.2.3. The Availability of Energy Supplies

Economists scorn geologists who depict energy supply as a function of time, as Hubbert did with oil. Political scientists, especially those with an

---

*In "Scenario B'," however, total demand would climb to 133 quads and energy for primary electrical generation would total approximately 45 quads. Such a scenario would require an annual GNP growth of 3 percent over 30 years. See Chapter Two for an explanation of why we believe this to be an unlikely event.

ear to the ground in Africa or the Middle East, discount the usefulness of pure economics as a guide to future energy supplies. It is with some trepidation, therefore, that we borrow the diligent work of economists and geologists in projecting the availability of energy for the U.S. in the year 2000 and report it here.

The energy market, even distorted by subsidies as it is, will determine consumers' choices of energy carriers. If alternate forms such as solar energy, electricity, and synfuels cost $10 to $20 per million BTU, then much conventional oil and natural gas can be economically produced before the alternates are competitive. The production of these fuels thus depends on competition, consumers' willingness to pay, and the distorting effects of subsidies.

Having made this caveat, we may return to Hubbert's projection reported in Chapter Four. Hubbert's curve indicates that 1.5 billion barrels of oil per year is all we should expect in the year 2000 from conventional U.S. oil production. Tertiary oil recovery could add one-half billion barrels per year. In BTU, the two sources would total about 13 quads per year. This is 27 quads of oil (12 million barrels per day) less than we presently use.

Natural gas is more of an unknown quantity. As we indicated in Chapter Four, a great deal of gas exists in deposits not associated with oil. For this reason, and because the price of gas is being deregulated, we believe that 20 quads of gas, perhaps more, could be produced in the year 2000.

It will be no easy task to double coal production by the end of the century. To do it safely and without a great deal of suffering, we will have to implement new (to the U.S.) mining technologies and ways of using coal. Maintaining the current level is, of course, feasible, but going from 17 quads per year to 34 should be considered a tenuous, if not undesirable, course.

Nuclear energy, strongly opposed by at least one-third of the U.S. population, widely perceived as a risky investment by financiers, and expensive in terms of delivered energy, will surely not contribute the almost 50 quads envisioned by USBM-1975. The 300 nuclear plants projected by iea for the year 2010 also seem very unlikely.

Wood waste, according to several sources, could supply up to 10 quads by the year 2000, while other sources of biomass could add five quads. The use of wood waste, however, must surmount some very difficult institutional and environmental problems. A biomass supply of 10 quads is optimistic, we feel.[5,6]

Solar energy, used actively, passively, directly, or to make electricity, and including presently existing hydroelectric sites and biomass, has been forecast variously to provide from zero to 20 quads of energy per year by the end of the century.[7,8] If biomass is included, and savings due to combined use of insulation and insolation are properly allocated, then solar energy could well supply 20 quads by the year 2000, but only if we pay its higher price and lay the groundwork now for its implementation. The key to making solar energy economical is to internalize to the fullest extent possible all the external costs of conventional energy supplies. These costs include national security threats, aesthetic degradation, ecological decline, and thousands of cases of human death and suffering from the health effects of energy use.

One point is reassuring, however. A demand level of up to 75 quads per year in the year 2000 can be met, though not easily. To build an economy on the presumption that substantially more energy will be available seems foolishly optimistic.

7.3. The Costs of Energy

Here we turn to summaries of projected costs of energy (see figure 7.1). These include the prices of the various energy carriers in terms of dollars per million BTU, and how this cost is affected by the thermodynamic efficiency of energy use. Related to thermodynamics and energy prices is the in-

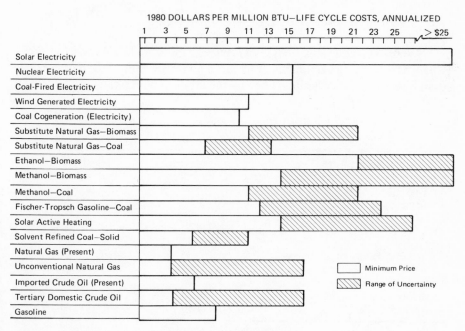

Figure 7.1. Synthetic energy carriers: A (delivered energy) cost comparison. Source: References 4–9.

tersubstitutability of energy carriers, a point which we discuss in this section. Costs also include the human health and environmental effects of energy production and use. By identifying those options with the lowest economic and environmental costs, we can suggest an energy supply policy. Such an evaluation requires summarizing the external costs of various energy supply options, making assumptions for energy production costs, and considering the magnitude, reliability, thermodynamic availability, and transitional difficulties of the various energy sources.

7.3.1. Present Energy Prices

The energy prices we will compare are prices of delivered energy, the prices of energy carriers in a form and place in which the consumer can use the energy. We present both average and marginal energy prices (see Table 7.2).

Note that the present delivered price of refined petroleum products is roughly twice the cost of crude oil. Refining adds 70 percent to the cost of crude; state and federal taxes and distribution costs represent the rest of the delivered price.

7.3.2. The Economic Costs of Alternate Fuels

What will alternate fuels cost? We can only make educated guesses. Chances are that the estimates of price we report in Table 7.3 are low rather than high, because inflation is hitting the labor and materials sector critical

Table 7.2 Average and Marginal Prices of Today's Energy Carriers
(1980 Dollars per Million BTU)

Energy carrier	Average price	Marginal price
Crude oil	$ 4.60	$ 6.50
Gasoline[a]	10.00	13.00
Diesel	9.00	12.00
Fuel oil	9.00	12.00
Heavy oil	3.50	4.70
Natural gas—industrial customers	3.00	5.00
Natural gas—residential customers	4.00	5.00
Coal—Electric utilities	1.25	1.20
Electricity—industrial customers	10.00	15.00
Electricity—residential customers	11.50	15.00

[a] Price includes state and federal taxes which combined average $0.14 per gallon.

[b] This is the cost to provide incremental power, which is usually not the same as the cost of new service to customers who would pay an average price.

Table 7.3. Estimates of the Cost of Alternate Energy Forms
(1980 Dollars per Million BTU)

Energy carrier	1990's cost	2000's cost
Nuclear–electricity	$17	$20
Coal–electricity	17	20
Coal–cogeneration–electricity	11	14
Wood–waste–cogeneration–electricity	10	14
Solar central–receiver–electricity	30	?
Solar photovoltaic–electricity	30–60	?
Wind–electricity	12	?
Solar passive–space conditioning	5–20	?
Solar active–space conditioning	15–25	18–30
Direct industrial use of coal	4	6
Fischer–Tropsch gasoline from coal	14	17
Coal to crude oil to gasoline	10–16	12–24
Oil shale to crude oil to gasoline	9–14	11–17
Biomass to substitute natural gas	10–18	12–27
Biomass to alcohol	8–15	?
Coal to alcohol	10–15	?
Natural gas	3–10	10–15
Coal to substitute natural gas	8–11	15
Oil	5–11	12–15

Source: References 5–9.

to energy particularly hard. Also, estimators tend to under rather than overestimate such costs.

Note in comparing Table 7.3 with Table 7.2 that almost every energy option will cost much more than we presently pay. Remember, too, that certain forms of energy are inherently more valuable than others.

7.3.3. Thermodynamics and Energy Policy

The energy born of the light and dust of a star is becoming degraded. The potential, kinetic, chemical, thermal, atomic, and/or nuclear energy contained in a body becomes less available as its energy is transferred in the form of heat or work to another body. When energy is transferred, thermodynamics becomes a useful tool if applied to minimize the loss of energy availability. It can be used to help match an energy carrier to a given task in order to make the best use of that energy and to reduce energy losses and costs.*

*For an exceptionally clear discussion of thermodynamics and energy policy, see Ralph Rotty, Reference 10.

The laws of thermodynamics state that

1) energy can neither be created nor destroyed; it can only be changed from one form into another.
2) energy flows only in one direction—from a hotter to a colder body—and that in this process entropy increases. That is, the availability of energy decreases or is degraded as it is transferred.

The first law states that the energy in the universe is always constant, but, according to the second law, whenever work is done or heat is exchanged, energy becomes more random and therefore less useful for doing work. A star may explode and spray its bits to the edges of the universe without energy (or matter) being destroyed. But the transfer of all that concentrated energy from the hot star to cooler matter (and the transfer of energy in the collision of star particles with other forms of matter in the universe) results in a dissipation of energy. The cool matter cannot give its newly received heat back to a hotter body. It can give heat only to a cooler body, and thus can only further dissipate it.

In practical applications of thermodynamics, within the usually overriding constraints of time and materials costs, we concern ourselves with getting the maximum possible work out of the chemical energy in the fuels we use. Energy conversion is most efficient when it takes place as slowly and with as much difference in temperature between the two bodies as possible. We are usually in far too great a hurry to do things slowly enough to achieve high efficiency. Also, the amount by which we may increase the highest temperature in a heat engine is limited by the ability of the materials in the engine to endure such heat. This ability, in turn, is a function of how much we can afford to spend on the engine. When energy was cheaper, the total cost of using it was minimized by using more energy but cheaper materials, or taking less time (labor) and using more energy. As the price of fuel increases, it becomes necessary to minimize costs by substituting more capital. This may mean using more efficient engines, larger heat exchangers, heat pumps, and other options which have been cheaper than fuel. But remember that substituting ingenuity for energy minimizes entropy.

There are two "figures of merit" which are useful for measuring the efficiency of energy use:

1) the effectiveness coefficient, which is the ratio of energy actually giving a desired effect to the energy consumed.
2) the ratio of the effectiveness coefficient to the maximum theoretical effectiveness coefficient.

The effectiveness coefficient is sometimes referred to as the first law efficiency, though, strictly speaking, its derivation or use owes nothing to thermodynamics. It is a particularly useful tool in comparing the efficiency of using similar energy forms to accomplish a given task. It is useful to know, for example, that one type of heat engine, the diesel, will extract more energy from oil products for delivery to the drivetrain of a car than will other types of heat engines such as the Otto cycle or regular auto engine. The diesel is more efficient because it operates at higher temperatures and because refining diesel fuel is more energy efficient than refining gasoline. But the effectiveness coefficients for these engines will not be useful when compared against an electric motor. Neither is the effectiveness coefficient particularly useful for comparing electric space heaters to gas furnaces for space heating. The issue is how well the energy supply and the energy demand are matched thermodynamically. Efficiency is best measured by comparing the effectiveness coefficient obtained with that which could have been obtained for a particular task. The ratio of the effectiveness coefficient to the maximum theoretical effectiveness coefficient is the thermodynamic, or second law efficiency. Exemplifying the application of these "figures of merit" to space heating, transportation, and industrial process heating should serve to illustrate the usefulness of thermodynamics as an energy policy tool.

7.3.3.1. Space Heating Efficiency

Comparison of a gas furnace with an electric heater for space heating demonstrates that the former will deliver about 70 percent of the energy in the gas to the living space (if the furnace is well serviced), while the latter will deliver virtually 100 percent. Does this mean that electricity is more efficient than gas? Of course not, since the 70 percent energy loss in generating and distributing electricity for the home has not been accounted for. Multiplying .3, the efficiency of electrical generation and transmission, times 1 (100 percent) yields .3, or 30 percent efficiency. Thus the effectiveness coefficient of gas, .7 or 70 percent, is far higher.

The maximum efficiency for heating a room, however, can be obtained with a heat pump.* The formula for obtaining this maximum value is shown in Table 7.4.

The formula, when used in a situation where outdoor (ambient) temperature is above 32 °F (273 °Kelvin) and the desired indoor temperature is 68 °F (293 °K), yields a coefficient of performance of ten. The coefficient of performance (COP) is the maximum effectiveness coefficient. Although it is far greater than one, it does not violate the laws of thermodynamics, for

---

*A heat pump is like a refrigerator; it cools the outdoor air but pumps the extracted heat into the living space.

Table 7.4. Calculating Thermodynamic Efficiencies[a]

	T_1 (hot)	T_2 (warm)	T_0 (ambient)	T_3 (cool)

System	A Effectiveness coefficient	B Theoretical maximum effectiveness coefficient[b]
Electric motor	$\dfrac{\text{Mechanical work out}}{\text{Electrical work in}}$	1
Electric heat pump	$\dfrac{\text{Heat added at } T_2}{\text{Electrical work in}}$	$\dfrac{1}{1-(T_0/T_2)}$
Electric refrigerator or air conditioner	$\dfrac{\text{Heat removed at } T_3}{\text{Electrical work in}}$	$\dfrac{1}{(T_0/T_3)^{-1}}$
Heat engine (including electric power station, combustion engine, etc.)	$\dfrac{\text{Electrical or mechanical work out}}{\text{Heat from hot source at } T_1}$	$1-(T_0/T_1)$
Electric heaters	$\dfrac{\text{Heat at } T_2 \text{ (or } T_1)}{\text{Electrical work in}} \approx 1.0$	$\dfrac{1}{1-(T_0/T_2)}$
Combustion heaters	$\dfrac{\text{Heat delivered at } T_2}{\text{Heat from hot source at } T_1}$	$\dfrac{1-(T_0/T_1)}{1-(T_0/T_2)}$
Heat powered refrigerator	$\dfrac{\text{Heat removed at } T_3}{\text{Heat from hot source at } T_1}$	$\dfrac{1-(T_0/T_1)}{(T_0/T_3)^{-1}}$

Source: Reference 10.

[a] The second law efficiency is the ratio of A/B.

[b] This is the maximum effectiveness coefficient which can be obtained in performing the given task, not necessarily with the same type of device. For example, the formula shown for the theoretical maximum shown for electric heaters is the same as that for heat pumps because an ideal heat pump is the most effective way of performing the task of delivering heat at an elevated temperature.

energy is not created, it is merely collected from the solar energy in the outdoor air. Entropy actually increases when energy is expended to run a heat pump. We can conceive how by considering that work would be required to reverse the process by cooling the living space and putting the heat back outdoors. There would be a net loss of energy.

To derive the ratio of the effectiveness coefficient to the maximum effectiveness coefficient, we take the overall values for the former which are .3 for electric generation and distribution and 2.5 for the actual effectiveness coefficient (COP) of today's heat pumps.* Then we divide by ten, the theoretical maximum for the aforementioned temperature range. The result, .3 × 2.5/10 equals .075, or 7.5 percent.

Now we may compare the effectiveness of an electric heat pump to that of a gas furnace. The gas furnace's effectiveness coefficient, .7, should also be divided by ten, the maximum value for this given task. The result, seven percent, is less than for the electric heat pump. Note two additional values, however. Electric heaters without heat pumps have a second law efficiency of 1 × .3/10 equals .03, or three percent. Gas furnaces using gas-fired heat pumps, which have effectiveness coefficients of two, can have a second law efficiency of 14 percent.

To achieve maximum efficiency in home heating tasks, we would always use heat pumps, unless even more efficient ways of heating homes could be conceived. The fuel cell concept, that of using gas to generate electricity on-site for use in buildings and applying the waste heat of generation for space heating, would be even more efficient. Using the waste heat from power plants in district heating systems also qualifies as highly efficient. Heat that otherwise would be discharged to the environment and eventually dissipated to outer space is put to use. Even individual households can cogenerate, using some of the electricity, exporting to the utility any excess, and applying the waste heat for space conditioning.

The point for policymakers is that total energy costs must be examined. Although electricity may cost twice as much as oil, it still might provide a better bargain if it does more than twice as much work per BTU. Unfortunately, the high cost of heat pumps is such that the average price of electricity at $11.00 per million BTU is effectively reduced to only $10.00 per million BTU when the cost of the heat pump is included. But this is far cheaper than the $15.00 marginal cost of electric power. In short, the better investment for society is more efficient equipment instead of new generating capacity. Situations which invite conservation through more efficient equipment like heat pumps and cogeneration plants provide the basis of advocates who propose that utilities offer subsidized loans for consumers to invest in on-site energy-saving equipment. It represents a better investment for both consumer and utility.

7.3.3.2. Automotive Efficiency

Automotive efficiency is related to many things such as the thermodynamic efficiency of the engine, the aerodynamic design of the car, the

---

*The actual effectiveness coefficient vs. the theoretical maximum efficiency is relatively low due to the high cost of materials, working fluids, and design for improving this value.

weight of the machine, and so forth. These are the subject of Chapter Nine; however, the efficiency of the engine is important for energy supply policy.

Table 7.4 shows how to derive the efficiency of automobile engines or of any heat engine. The theoretical maximum efficiency, which is a function of the maximum temperature an engine's materials can withstand, is about 87 percent. Otto cycle (gasoline) engines, however, obtain at best 10 percent thermodynamic efficiency today. Electric automobiles, with advanced batteries, could obtain 20 percent efficiency when we include power plant and transmission losses. Should we thus convert to electric cars? Although the cost of electricity is about twice that of gasoline, the electric engine would be twice as efficient and therefore no more expensive in terms of fuel costs. But the costs of manufacturing electric automobiles would be higher than that of producing conventional cars. The automotive industry, the largest industry in the world, has enormous amounts of capital invested in making certain types of cars. What would be the cost of scrapping this equipment and starting over? And would consumers buy low performance, short-range automobiles such as electric cars? Answers to such questions, not thermodynamic efficiency alone, will dominate the future of the automobile. Still policymakers have an obligation to be aware of the implications of their choices in sponsoring certain systems for research and development and/or subsidy. The overall thermodynamic efficiency of a gasoline engine fueled with gasoline made from coal via the Fischer–Tropsch process would be only one-third as efficient as today's cars: three percent efficient as opposed to ten percent today. It is clear that dramatically increasing the poor efficiency of gasoline engines is our highest priority.

7.3.3.3. Industrial Energy Efficiency

In relation to the other sectors of the economy, the efficiency of energy use in industry is quite high. As Table 7.5 shows, industrial process steam and direct heat applications are 25 and 30 percent thermodynamically efficient, respectively. Industrial cogeneration, a subject we discuss at length in Chapter Ten, can be 50 percent efficient. Usually in industrial cogeneration, more electricity is generated than can be used on-site. This power can be exported for other users when institutional problems are overcome. And usually where there are large industrial steam applications, cogeneration is economical. Since the costs of steam generation may be allocated between industrial and off-site electricity users, the costs to both the industry and the affected user go down.

Table 7.5. The Thermodynamic Efficiency of Providing Major
Energy Services

Service Applications	Fraction of total U.S. fuel consumption	Estimated overall thermodynamic efficiency (percent)
Space heating	18	6
Water heating	4	3
Air conditioning	2.5	5
Refrigeration	2	4
Industrial uses		
Process steam	17	25
Direct heat	11	30
Electric drive	8	30
Transportation		
Automobile	15	10
Truck	5	10

Source: Reference 10.

In scenarios such as the iea–1978, however, the idea seems to be to generate power in large, central station sites, and use electricity for every conceivable purpose. We might ask what the thermodynamic and energy costs of such an approach might be.

For industrial steam and direct heat applications, the thermodynamic efficiency of first generating steam to make electricity (which is then converted back to steam or hot air by heating water or air with electric coils) is simply that of any heat engine. Theoretically, the effectiveness coefficient, using present-day materials, is 85 percent, but in reality it is only about 30 percent. Multiplying .3 by .9, to account for the slight inefficiency of converting electricity into steam, hot water, or air, yields .27. Dividing by .85 gives an efficiency of 32 percent. While this rate is comparable with present practice, it is far below the potential efficiency of industrial cogeneration, and below as well what will be achieved as industry responds to energy price increases. Furthermore, deriving work from energy that costs $15.00 per million BTU instead of $3.00 (and cost only $0.50 per million BTU when the plants were built), is an extraordinary increase. Electric energy's application to industrial processes will be avoided because of its great cost.

Table 7.5 summarizes the efficiencies of major applications of energy in the U.S. Those market segments with the lowest efficiencies and the highest percentage of demand offer the greatest technical promise for reducing energy demand.

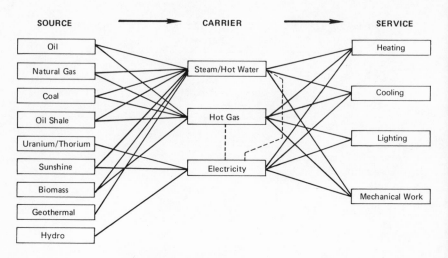

Figure 7.2. Flexibility in providing energy services.

7.3.4. Energy Carrier Interchangeability

We noted in Chapter Three that industrial nations possess a great deal of flexibility in the selection and use of energy carriers. In Figure 7.2 we demonstrate that point by illustrating the various paths between energy sources and energy services. The basic energy services provided are heat, cooling, light, and mechanical energy. Heat may be provided by any of the energy carriers, be it steam, hot combustion gases, or electricity. Cooling may be derived from steam (or hot gas) absorption compressors, or electric refrigeration. Light may be provided electrically or by the luminescence of hot combustion gases. Mechanical work may be derived from the expansion of hot gases or steam, or from an electric motor. Steam and hot gases may be used to produce electricity and vice versa. Most energy sources can be converted to steam, hot gas, or electricity. Even those that cannot be directly converted to steam, for example, produce electricity which then can be converted to steam.

What, then, does thermodynamics tell us? It describes our flexibility in providing energy amenities with our various indigenous energy sources. It indicates where and sometimes how we can make the most effective use of energy. But thermodynamics cannot be the basis for choosing one path over another. For, inevitably, it will be left to economics to minimize the cost of several of society's resources, including time, money, national security, and environmental health. The clearest signal that consumers can receive on the value of each of these "goods" is the price of goods and services.

7.4. The External Costs of Energy

It is a fact that many thousands of people die each year from our use of energy. Some analysts estimate that the cost of using coal for making electricity alone costs 20,000 lives per year. It is this statistic that spurred many nuclear advocates to the belief that it was worth taking the risk of a catastrophic nuclear accident in order to prevent the high human cost of using coal. That cost makes the more efficient use of electricity, regardless of the source, not only economically sound, but morally imperative. In Chapter Six we presented the case for including the external costs of energy use in the price the consumer sees by requiring, for example, cost-effective pollution control. Far from being a luxury, buying pollution control is inherently rational.

The need for pollution control is indicated in the relative damage of the various fuel-to-electric systems. As Table 7.6 indicates, the present cost (estimated for 1974) of using electricity is quite high. Coal-fired electrical generation annually takes the lives of 2000 to 20,000 and disables 20,000 to 40,000; oil-fired electrical generation kills from 100 to 5000 annually and disables 4000 to 9000; nuclear and natural gas cause far less harm, though nuclear energy use carries with it even greater risks. Using published data from the National Academy of Sciences, we projected the year 2000 level of insult for the USBM-1975 and iea-1978 scenarios. They show levels of

Table 7.6. Estimated Annual Health Effects of Electrical Generation in Selected Scenarios

Scenario	Deaths[a]	Disabilities[a]	Unknowns
1974 actual (estimated)			
Coal	2000–20,000	20,000–40,000	Acute and chronic diseases;
Nuclear	10	500	carcinogenic and genetic
Gas	10	600	effects
Oil	100–5000	4000–9000	
USBM–1975 for year 2000			
Coal	800–900	19,500–25,500	Acute and chronic diseases;
Nuclear	230	11,000	carcinogenic and genetic
Gas	10	300	effects, plus catastrophic
Oil	30	2600	accident potential
iea–1978 for year 2000			
Coal	900–1100	17,700–29,000	Acute and chronic diseases;
Nuclear	80	3700	carcinogenic and genetic
Gas	—	—	effects, plus catastrophic
Oil	40	3600	accident potential

Source: Reference 11.

[a]These numbers are not normalized to a unit of production.

death and disability comparable both to the present and to each other in the year 2000. Although energy demand is far higher in the USBM-1975, the percentage reliance on electricity is much greater in the iea-1978 projection.* Thus the two are comparable in forecasting the health effects of electrical generation. Projections of the effects of other energy activities for the two would differ. The fact that both show levels of insult comparable to the present despite the far greater use of electricity reflects the assumption that antipollution laws will be enforced. Since there is reason to believe that our leaders may capitulate on this issue, the death toll for a given level of electrical usage and for all energy use could go far higher.

Natural gas comes as close as anything we have to being environmentally benign. Little land is disturbed in its production, transmission in underground pipelines is inoffensive, except on rare occasions when they explode, and conversion yields mainly carbon dioxide and water. Only the carbon dioxide produced in the combustion of natural gas presents any real, long-term threat. In the event that the carbon dioxide problem forced us to abandon fossil fuels, natural gas could have a surrogate in methane gas derived from solar biomass.

Biomass could be benign. But careless management and rapid exploitation could deplete and destroy our crop and forest lands, a catastrophic prospect.[6] Like most technologies, biomass energy systems could be benign if handled properly. Generally, however, the failure to manage properly bio–energy systems would be one threat that could be reversed. The failure would manifest itself over a period of time, and time enough to correct the problems should exist. For biomass to be a renewable resource, its use must be carefully husbanded. Thus, in the long term it must be benign; in the short term when use is low, it will be benign. The problem will be in the transition.

Direct solar energy systems are not entirely benign, because the production of collectors, silicon cells, trombe walls, etc., requires energy for the fabrication of materials. But since production is a one-time process, and environmental insults are not a function of the operation of solar systems over their lifetimes, direct solar systems may be considered benign.

Oil, unconventional natural gas, synfuels from biomass in the midterm, high efficiency use of coal, geothermal, central solar electric, and hydroelectric systems are energy sources which may produce locally catastrophic problems, but problems which can be contained. Oil spills can be devastating locally. Tertiary oil and gas recovery may require a great deal of steam generation and associated air pollution. Biomass may be overexploited and ecosystems and aesthetic values may suffer.[6] Coal and geothermal systems may produce serious air and water pollution and land

---

*Recall that the iea-1978 scenario assumed a nuclear moratorium to begin in 1985.

disturbance. Hydroelectric dams may systematically eradicate fresh-water crustaceans and fish. But all of these, with the exception of the potential global CO_2 problem with fossil fuels, could be managed properly.

Systems which could lead to such serious environmental problems as to disqualify them from use include both coal and nuclear central station electric generation, imported oil, and synthetic fuels from coal and oil shale (see Chapter Six for a detailed analysis of the environmental impacts of these systems).

7.5. Future Electric Demand

It has been the *de facto* energy policy of the U.S. for thirty years to prepare to substitute electric energy for diminishing supplies of oil and natural gas. But the reality of high electric costs, both implicit and external, the potential for higher power plant capacity utilization and for conserving electric energy at the point of use, and the unsuitability of electric energy for many applications, makes the forecasts of an all-electric future seem passé.

It is not naive to ask the question, "What would one do with more electricity?" If we answer that we will use additional electric energy to heat buildings, then we overlook the opportunity to cut energy demand in the buildings sector by 25 to 50 percent, as well as the improved outlook for gas availability. If we reply that we will substitute electricity for the oil imported for automotive uses, then we ignore the extreme difficulties of electrifying the automotive fleet which would mean completely retooling Detroit to produce cars with poor performance, short-range, high costs, and poor energy efficiency. Automobile petroleum use, if one is willing to accept in gasoline powered cars the performance of electric cars, can be slashed by 75 percent. If we respond that we will use more electricity in industry, then we fail to observe that substituting electric energy for natural gas imposes a five-fold increase in industrial energy costs, and that most industrial energy demand is for steam or low grade heat. The greatest use of energy in an electric energy future would be to make electricity.

To be sure, electricity is an invaluable tool in an industrial society. But its high cost and poor substitutability for liquid and gaseous fuels makes electricity a poor candidate for the chief energy carrier of the future. The high cost of building new electric generating capacity is so great that electric demand growth will be diminished to the extent that utility load management techniques, industrial cogeneration, and power plants under construction can saturate electric energy demand for the forseeable future.

Even the replacement of oil burning power plants is not an urgent matter. Most oil burning plants consume residual oil which comes from two sources. It is a natural by-product of refineries and compares in quality to asphalt. Until refineries are retrofit to synthesize higher grade fuels from residual oil, a step which may cost tens of billions of dollars and a decade or longer, the use of domestically produced residual oil is rational. Imported residual oil comes predominantly from Venezuela, is derived from low grade crude oil of secondary importance in the world oil market, and costs one-half to two-thirds as much as high-quality, high-priced, tenuously secured oil from the Middle East. Clearly, public subsidies to reduce oil imports should be applied elsewhere before low quality, secure, residual oil is "backed out" of electric generating plants.

The use of natural gas in industrial cogeneration, however, offers a more efficient use of this precious fuel than residential space heating. Electricity cogenerated by burning gas in industry could produce excess power for heating homes efficiently with electric heat pumps. Cogenerated electricity is a most efficient energy process.

7.6. Energy Pricing and Equity

Energy price controls garble the economy, help plunder our natural resources, forfeit our national security, protect the rich but inevitably crush the poor. As Senator Paul Tsongas observed, ". . . to protect the consumer from the reality of the energy crisis is to destroy him." Price controls along with tax subsidies for energy production and consumption serve primarily the middle and upper income groups who consume the most energy per capita. Energy price disjunctures ultimately erupt when energy prices are held artificially low for too long. When vastly more expensive energy supplies must then be introduced, wealth is redistributed from the poor to the relatively wealthy through the mechanism of inflation. If equity is protecting the poor, equity is badly served by energy price controls. Far better mechanisms exist for protecting poor people.

The impact of price controls on gasoline illustrates the general inequity of energy pricing policies. Gasoline consumption is income-elastic; that is, the higher one's income, the greater one's consumption of gasoline. Upper income families consume an average of 1000 gallons of gasoline per year, compared to about 400 gallons per year for low income groups. If we assumed that the unregulated price of gasoline would be $0.50 per gallon higher, then the gross subsidy to the wealthy family would be $500 per year, contrasted with $200 per year for the low income family. The distortion of the gasoline market, moreover, manifests itself in higher foreign oil imports which lead to a balance of payments deficit, and, as a result, inflation. The wealthy consumer can of course protect him or herself from inflation in a myriad of ways. Clearly, holding gasoline prices below market value makes the rich richer and the poor poorer. This is true to an inestimable degree when our scope is widened to include the Pakistani barber (see Chapter Three) who, by American motorists, is priced out of the market for kerosene for use for the most basic needs.

Control of the price of natural gas presents a similar dilemma. The argument is often made that the cost of producing a given quantity of natural gas is, for example, only \$3.00 per million BTU and, therefore, its price should be \$3.00 plus a reasonable rate of return on the producers' investment. But gas can substitute for oil which costs \$10.00 per million BTU or electricity at \$15.00 per million BTU. If the consumer invested in conservation options at only \$5.00 per million BTU, he or she could reduce household gas consumption for space heating by 50 percent. But with natural gas costing only \$3.00 per million BTU, it would be irrational for the industrial consumer to invest \$5.00 per million BTU in energy conservation, much less \$10.00 or \$15.00 per million BTU, the marginal price of the fuel or electricity which natural gas could replace (or which replaces natural gas when it becomes unavailable at low prices). The richer homeowner, who generally owns a bigger house and thus consumes more gas, is again subsidized by a greater amount. The economic inefficiency which results from the failure to save the real value of the gas results in inflation as alternative energy supplies are delivered to fill the lost conservation opportunity. Alternative supplies may include increased oil imports, harder to get and therefore more expensive natural gas, electric, or solar energy. Each of these will cost more than the conservation option and will thus be inflationary. Those options significantly more expensive than natural gas including electricity, synthetic gas, and solar energy, may be utterly out of reach of the poor.

It is also clear that missed energy conservation opportunities are missed chances to save lives routinely lost in the production and conversion of energy. When electricity is saved, fewer miners die mining the coal for electric generation and fewer cases of lung disease are caused by the effluents of electric power plants. Miners often express love for their work, but just the same, the fewer we send to dig coal, the fewer we send to die. It is often true, too, that the poorest segments of our society suffer the greatest damage from the effects of air pollution and other environmental insults. Increased energy consumption due to artificial energy price controls increases the suffering of the poor who are least able to avoid the harmful effects of energy use.

The total U.S. government subsidy of energy consumption due to energy price controls alone may exceed \$100 billion annually. Other subsidies such as the energy expense tax deductions, foreign energy development incentives, incentives for inefficient truck freight transport, etc., may surpass \$200 billion per year, or 10 percent of the GNP. For contrast, the annual \$1 billion Department of Energy expenditure on conservation programs appears pathetic. Tax credits for residential retrofits amount to no more than an annual $2 to $3 billion. The federal government, in fact, can never hope to offset the incentives for energy consumption that energy price controls furnish. To offer incentives for energy conservation of the magnitude required to ameliorate the effect of price controls and other energy consumption subsidies would cost one-third of the federal budget.

Equity, that is, protecting the poor and powerless from the impact of high energy costs, can be accomplished much more effectively with mechanisms for direct income redistribution than by energy price controls. In fact, equity cannot be served by energy price controls.

7.7. Toward Consensus

What remains is to organize this unwieldy set of information about energy options—economics, environment, physics, supply potential—into a framework from which transition strategies might suggest themselves. That is not to say we could plan to endure an all out oil embargo (coming before 1990) for very long without curtailing our activities, but we can envision an energy future which works. And one that offers us a way to survive such an embargo.

In the Prologue, we drew a picture of the overlapping waves of historical energy use in the U.S. Wood was overtaken by coal which was in turn overtaken by oil and gas. We noted that for three decades or more analysts have been drawing the next wave as nuclear. Now analysts are drawing a number of future waves, some of which are the mirror image of historical transitions. Coal will follow oil, and after oil will come wood or some other form of biomass and/or solar energy. Some worry that "we need every source of energy we can get" and draw composites of the various supply options in order to meet energy demand. The fact that each of the supply options carries with it great economic, human, and environmental costs is what drove us to turn to the potential for energy conservation as an energy option.

One can take the estimates about individual options and add them up in a number of ways. One can minimize economic costs by picking options with the lowest energy price (in dollars per million BTU) without regard to external costs. But in doing so, say, in order to add up enough energy supply to meet a demand level of 70–100 quads in the 1990s, one inevitably is faced with the choice of whether to buy imported oil or to go to some other source, such as shale gasoline. Shale gasoline, however, could supply only a very few quadrillion BTU, could cost twice as much as imported oil, and would rank high in external costs primarily because of the environmental and health risks associated with its production and refining.

Note that in meeting the lowest 1990s demand level, 75 quads, the following supply items might be relied upon for a total of 60–65 quads:

- o Natural gas, 20 quads
- o Wood waste, 5 quads
- o Domestic oil, 10 to 15 quads
- o Hydroelectric power, 2 quads
- o Solar energy passive applications, 2 quads
- o Coal–electric, 12 quads
- o Direct coal, 5–10 quads
- o Nuclear–electric, 4 quads

At a minimum, an additional 5 quads of energy would have to be: (a) imported; (b) produced from the direct use of coal; (c) added by nuclear electric power generation; (d) obtained by using solar energy; (e) produced as liquid or gaseous fuels from coal or oil shale; or (f) derived from further nonmarket conservation. Importing five to ten quads of oil in the 1990s, one-quarter to one-half the oil we now import, may be politically acceptable; it also may *not* be. Ten quads of coal could be produced by mining an additional half billion tons of coal each year, and the cheapest way to use it would be directly, especially in industrial cogeneration in industry. Manufacturing synthetic fuels such as gasoline with the Fischer–Tropsch process would require two to three times as much coal and would cost perhaps twice as much economically and environmentally. Nuclear- and/or coal–electric generation in large, central station power plants are options which cost far more than alternatives. Solar energy active heating and electric generating systems will carry high economic price tags, if not environmental costs. Synthetic fuels from coal and oil shale, like central station electric generation, should be last resorts. Their use will be constrained for at least the rest of the century to small quantities of energy production simply because of the physical problems inherent in synthetic fuels production. *Our main point, here, is that if supplying 70 quads will be difficult without compromising our quality of life, then producing 90 quads or more will be a grim effort.* Another major point, the point to which all of Part II has been leading, is that the natural effect of rising energy prices (and internalized costs) will be to lower energy demand. The technology and policy options for assuring that a desired amenity level can, nevertheless, be produced through energy conservation techniques is the subject of Part III.

Similar scenarios can be drawn for the years after the turn of the century and for the indefinite future. A major point of interest arises, though in the possibility that we may be able to lessen the pain of energy transitions by making a conscious effort now to strive for a truly long-term sustainable energy economy. The transition could be orderly, and, in the long run, such an energy economy will prove the least expensive. The future could be arrived at through a methane scenario or any other that researchers and the public may devise. In devising such a future, experts should aim research and development efforts at those options which minimize both direct and external costs. At the same time, it should be a matter of continuing public debate regarding what external costs are unacceptable, and when we should pay a higher price for energy in order to have a higher quality of life.

Conservation . . . public policy must be its champion.
—ROBERT STOBAUGH AND DANIEL YERGIN
in *Energy Future* (1979)

Part Three

The Conservation Well

Introduction to Part Three

A workable energy future would be one in which ingenuity was substituted for resources, in which technological and social innovation produced a higher level of amenities with less trauma, and one that allowed personal growth and freedom, but was sustainable. Such a future will require keen insight and great resourcefulness to generate steady growth in energy productivity—but it *is* achievable.

In Part One, we reviewed a number of studies of the future of energy demand and found that the higher estimates were derived from expectation of continued cheap energy and from a lack of attention to certain saturation effects and demographic changes. In Part Two, we examined evidence that the external costs of energy are already high (and not always internalized), and that the overall price of energy is likely at least to double over the next few decades. In Part Three, we attend to the factors of demand and focus on conservation as a resource. We consider cost-effective options for cutting buldings energy consumption in half, for halving the use of oil in automobiles, and for increasing energy productivity in industry so that the economy can grow though energy demand may not. We find that the problems are as much institutional as technical, and that a major commitment and great effort must be made to achieve the potential for energy conservation.

Our nation's residential and commercial buildings are manifestly overlit, overheated, overcooled, and underinsulated.
                                                        —MAXINE SAVITZ AND ERIC HIRST
                                in *Energy Conservation and Public Policy,* edited by John C. Sawhill (1979)

Chapter Eight

Buildings—More Amenities, Less Energy

8.1. A Myth

One of the most persistent myths about energy conservation is the one about thermostat setbacks. If one wanted a house or office building warm at eight a.m., according to folklore, it would save energy to keep the building warm all night rather than to let it cool down and then warm it up quickly at 7:30 a.m. But putting heat in a room is like putting water in a bucket. The longer a leaky bucket must hold water, the more it will lose. Likewise, the longer that a room must hold warm air, the more it will lose.

As a matter of policy, however, no one would advocate that home or building owners turn off the heat in their leaky buildings. Rather, it is the leaks that demand attention. That we must curtail our use of the amenities of energy, that we must do without, is yet another persistent myth, but one far more insidious.

Table 8.1. Energy Use in Buildings (1975)

End use	Percent of total energy use in buildings	Quads
Space heating	65	10.6
Water heating	11	1.8
Air conditioning	7	1.1
Lighting	6	.9
Cooking	3	.5
Refrigeration	2	.3
Food freezing	.6	.1
Other	6	1.0
Total	100	16.3 (23.4)[a]

Source: Reference 1.

[a]The lower figure represents consumer energy use; the larger equals total energy use. The former is calculated using electricity at 3413 BTU per kilowatt hour, and the latter was derived using a heat rate for the generation of electricity of 10,000 BTU/kilowatt hour.

8.2. Amenities

Increasing the effectiveness of energy used for space heating is the most important energy conservation option in the buildings sector. Of the amenities requiring energy in residential and commercial buildings, space heating consumes the greatest share, nearly two-thirds of 16 quads burned annually. Seven end uses of energy in the buildings sector account for 20 percent of the energy used in the U.S. In order of importance, these are space heating, water heating, air conditioning, lighting, cooking, refrigeration, and food freezing (see Table 8.1).

Obviously, the electric toothbrush is not a high policy priority. And we emphatically reiterate that neither is personal sacrifice. Although nighttime thermostat setbacks and reduction of water heater settings are behavioral changes which offer substantial savings, they are neither significant lifestyle changes nor the cornerstone of an energy conservation policy for buildings. Indeed, in the year 2000 we can have 50 percent more residences and 65 percent more commercial space, all comfortably heated and cooled and equipped with appliances to a degree approaching saturation and yet use 20 percent less energy than we do today. We would not be getting something for nothing, though. The scenario does assume that energy prices go up and that consumers respond by investing in conservation measures to save money.

In this chapter we assess the potential for saving money by saving energy in each of the major energy uses in both new and existing buildings.

We note areas where policy actions are required to help obtain the optimal solution, and also complexities caused by climatic, construction, and energy price differences.

8.3. Space Heating

8.3.1. The Thermal Integrity of New Homes

It is economical in the new houses of most regions of the country to reduce space heating energy demand to two-thirds that of the average new home. To illustrate the validity of this claim, we examine the cost of conservation options (Hutchins and Hirst[2]) for a new house in Kansas City (see Table 8.2 and Figure 8.1).

There are sixteen options for incremental conservation investment listed in Table 8.2 and which are graphically depicted in Figure 8.1. The first option is the baseline, a new Kansas City home which, virtually uninsulated, would consume 83 million BTU per year for space heating. Adding attic

Table 8.2. Conservation Features in a New Kansas City Single-Family Home

| | | | Cumulative additional initial cost (1980 dollars) | |
| Conservation option | | Percent of annual baseline heating load | Heating system only | Heating plus air conditioning |
Number	Description			
1.	Baseline (demand = 83 million BTU per year)	100	$ 0	$ 0
2.	Attic insulation R-11	76	180	2
3.	Wall insulation R-11	63	365	80
4.	Attic insulation R-19	59	460	130
5.	Floor insulation R-11	51	715	410
6.	Storm windows	45	910	590
7.	Attic insulation R-30	43	1075	720
8.	Floor insulation R-19	41	1180	845
9.	Double-paned sliding glass door	38	1350	1000
10.	Wall insulation R-13	37	1410	1065
11.	Attic insulation R-38	36	1520	1160
12.	Wall insulation R-19	34	1785	1410
13.	Triple-paned windows	32	2020	1635
14.	Attic insulation R-49	31	2175	1780
15.	Storm door	30	2320	1930
16.	Wall insulation R-23	29	2590	2195

Source: Reference 2.

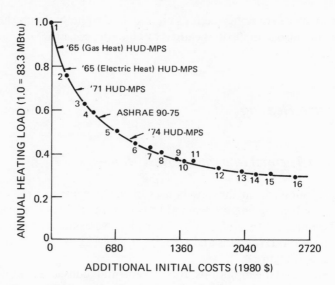

Figure 8.1. The cost of reducing residential heating loads in new homes: Kansas City example.
Source: Reference 2.

insulation with an *R* value (heat resistivity index) of 11 would reduce the baseline heating load to 76 percent. The additional *net* cost, indicated in column three, would total $180 (1980 dollars), if air conditioning is not installed. If air conditioning is added, the incremental cost is only $2. The $178 difference would be saved by virtue of the fact that the insulation would reduce the size of the air conditioning unit required. The additional cost shown in the "heating system only" column reflects a similar credit. Here we have an example of a capital investment for insulation, substituting not only for fuel for heating and cooling, but also for expensive energy-consuming equipment (the otherwise larger heating and cooling system).

Moving down the list of options, which are cumulative in both energy reduction and cost, to, for example, option 13, we find that a total extra investment of $2020 in the new home will reduce the heating load of that house by 68 percent.

The economic optimum for any given home will depend on a variety of factors such as the local climate, choice of energy source, desired style of building, and so forth. An analysis by Hirst and Kurish[3] based on fuel choice and climate indicated a range of optimum solutions. A new home builder in Miami using gas heat and air conditioning should, according to the assumptions made by Hirst and Kurish, invest in options one through four. The optimum investment for Minneapolis, however, would be in all options 1–16, if the new house were electrically heated and cooled. Table

8.3 relates the conclusion of the above analysis for nine different cities. The table also compares the economic optima with the ASHRAE 90-75 standard, which is promoted by the American Society of Heating, Refrigeration, and Air Conditioning Engineers, and incorporated into the building codes of 45 states. Note how much lower is the ASHRAE standard in effect at this writing. The only option called for in Miami is number four, and in Minneapolis number six. The issue of what the building standard should be is one that is likely to remain controversial for many years. We return to this debate in Section 8.3.2.

But first we should consider two very important assumptions that entered Hutchins and Hirst's analysis. The optimum levels of investment they suggested were based on (1) the assumption that the conservation investment would be amortized over the life of the investment in the home (typically, a 30 year mortgage), and (2) that energy prices would not increase in real terms beyond 1976 levels. The first assumption either increases or decreases the indicated optimum level, depending on one's frame of reference, while the second assumption definitely decreases the indicated optimum.

One could argue that homes last longer than 30 years, that conservation investments will thus deliver energy savings far longer than the 30-year life of a mortgage, and therefore that such investments should not be discounted so heavily. More commonly, one hears the argument that home builders should not be burdened with investments on which they will not gain returns. Since most people own homes for only about seven years, the

Table 8.3. Conservation Features for New Homes: Economic Optimum vs. ASHRAE 90-75 Standard[a]

Location of new home	ASHRAE 90-75 standard[b]	Economic optimum with 1976 prices	
		Gas heat and air conditioning	Electric heat and air conditioning
Minneapolis	6	11	16
Boston	4	9	14
Seattle	9	9	14
Kansas City	4	7	14
Washington, D.C.	5	7	13
Atlanta	4	4	12
Fort Worth	3	4	11
Phoenix	3	4	9
Miami	2	4	4

Source: Reference 3.

[a] Refer to Table 8.2 for description of the options.

[b] ASHRAE makes no recommendations regarding the economic factors relative to different fuel sources.

argument goes, the payback period on conservation investments should not exceed seven years. This argument, however, ignores the increase in resale value an energy conservation investment creates.

Further, none of the indicated optima in Table 8.3 has a payback period greater than six years, even without the real energy price increases that have occurred since 1976. The Minneapolis investment cases, for example, would have payback periods of two to four years, periods similar to the Kansas City cases and to most others.

Another way of looking at residential energy conservation investments is to compare their cost in terms of dollars per million BTU saved with the cost of producing marginal, or new, sources of energy (Table 8.4). This perspective differs from that reflected in Table 8.3 in two ways. First, the recent increases in energy price are incorporated, and second, the perspective of the cost to society, rather than to the individual consumer, is taken. Despite the fact that a new consumer gets energy from an electric or gas utility for an average price, the marginal, or incremental cost of that energy is higher. This higher cost is borne by all of the utility's customers. Thus, the new customer is subsidized by society, that is, by the other consumers whose average costs are driven higher by the addition of the new customer's incremental cost.

The chief point to be gained from Table 8.4 is that it is more economical for society and/or a utility to subsidize or otherwise provide incentives for energy consumers to invest in energy conservation. This rationale lay behind the investment tax credits for investments in conservation and solar energy proposed by President Carter in his National Energy Plan, which Congress approved as the National Energy Act/Energy Tax Act.[4] Until January 1, 1986, 15 percent of every investment up to $2000 in residential energy conservation will be returned as a (non-refundable) tax credit. A home builder investing in option 16 would thus get a tax credit of $300.

Homebuilders are not investing in the optimum energy conservation options. This failure is due in part to the desire (or necessity) to minimize initial investment costs, to lack of information, and/or lack of incentive due to average cost energy pricing or building practices in which the persons who pay the heat bills do not make the initial investment decisions. A number of federal efforts have been tried as ways of circumventing these problems. These have been less than successful.

8.3.2. Thermal Standards for New Homes

Federal efforts to encourage energy conservation in buildings include three major tools. The Department of Housing and Urban Development

Table 8.4. New Energy Supply Costs vs. Conservation Costs in New Homes
(1980 Dollars per Million BTU[a])

Location of new home	New supply costs		Selected conservation investment costs	
	Gas	Electricity	Option 13[b]	Option 16[b]
Minneapolis	$3.00–$6.00	$10.00–$15.00	$ 1.35	$ 1.82
Boston	3.00– 6.00	10.00– 15.00	1.85	2.45
Seattle	3.00– 6.00	10.00– 15.00	1.70	2.30
Kansas City	3.00– 6.00	10.00– 15.00	1.95	2.55
Washington, D.C.	3.00– 6.00	10.00– 15.00	1.90	2.70
Atlanta	3.00– 6.00	10.00– 15.00	2.85	.3.85
Fort Worth	3.00– 6.00	10.00– 15.00	3.50	4.75
Phoenix	3.00– 6.00	10.00– 15.00	2.40	3.25
Miami	3.00– 6.00	10.00– 15.00	10.75	14.50

Source: Reference 3.

[a]Annualized, life-cycle costs.   [b]Refer to Table 8.2.

has a set of "Minimum Property Standards" (HUD-MPS) that apply to all homes with financing provided, insured, or guaranteed by the federal government. These standards apply to about 25 percent of the new home market, and, until 1979, when they were improved, were only slightly more stringent than ASHRAE 90-75. The rest of the housing market, built during the latter half of the 1970s, was constructed, generally, according to ASHRAE 90-75. The ASHRAE standards were incorporated into state building codes in response to the Energy Policy and Conservation Act of 1975 (P.L. 94-163), which required states to adopt minimum mandatory thermal efficiency standards for new home construction in order to receive federal funds for energy conservation purposes. ASHRAE 90-75 is considerably less stringent than proposed new standards required by the 1976 Energy Conservation and Production Act (P.L. 94-385). The new standards are called Building Energy Performance Standards (BEPS) and differ in philosophy from ASHRAE and HUD-MPS. BEPS, for instance, requires that buildings actually perform at certain levels of thermal efficiency, rather than simply be built with certain types and quantities of energy conserving materials.

BEPS has been compared to the minimum mileage efficiency standards required for new automobiles.[5] Like those standards, BEPS has come under heavy fire. BEPS, unfortunately, is plagued by considerably greater complexity and by the problem of enforcement. The variety of house styles, fuel prices, climates, and so on, caused authors of BEPS to resort to a complex calculus for the construction industry to follow to determine how to build a structure which complies. A danger is that the federal government will not enforce BEPS, and that the far weaker ASHRAE 90-75, or some revised but still inadequate surrogate, will substitute for it. If this substitution comes to pass, a great opportunity will have been missed, because BEPS is estimated to be capable of saving the equivalent of up to a quarter of a million barrels of oil per day, or .6 quads per year, by 1985.[6]

8.3.3. Passive Solar Energy Systems for Conservation

Demand for space heating varies from less than 10 million BTU per year per household in south Florida, to more than 200 million BTU annually in Minnesota.* Passive solar energy systems such as a sunspace or greenhouse attached to a home have been tested and found to deliver up to 85 million BTU of space heat per year in sunny locations. More typically, a greenhouse delivers about 25 million BTU of heat annually, the equivalent of four barrels of oil. The additional construction cost, the equivalent of $5 per million BTU, is half as expensive as heating oil. It is, therefore, economical for homes in many parts of the country, particularly in the southwest, to derive almost all of their heat from passive solar energy systems. A very well insulated Kansas City home could obtain 75 percent of its space heating needs from passive solar energy systems at costs currently competitive with oil. This level of performance would require a $1000 investment for the passive solar features in addition to $2700 for thermal integrity. A much larger investment in other passive energy systems, perhaps $8000 for a thermal storage wall, would provide 40 million BTU annually at a cost of $10 per million BTU, a cost cheaper than heat from an electric heat pump. Still, other means would be required to furnish backup energy supply to the Kansas City home, a factor that would drive up the cost. Back-up costs should be counted as a cost of solar energy.

Solar greenhouses are the cheapest of the passive solar energy options which include both Trombe walls and water walls for indirect heat gain and storage and solar roof ponds for direct energy collection. Assigning costs to solar energy passive devices is difficult, however, for many of the gains can be had with simple building design changes, such as placing more window area on the south side. Also, passive solar energy applications in the home are frequently attractive and provide amenities in addition to heat, and these other amenities might share the extra cost.[7]

The issue of backup requirements has been raised as an objection to solar energy systems in general. The objection is that solar energy systems would strain conventional utilities by creating an excess demand load on days the sun did not shine. This problem has also been described as

---

* This level of energy use is not directly comparable to the Kansas City home example in Figure 8.1, because the Kansas City example assumes only *commercially* purchased energy and ignores appliance waste heat and solar passive energy gain.

overrated, with analysts pointing out that a high growth rate in the use of solar energy systems would be required before utilities would even notice such loads. When such loads do become larger (assuming that backup energy were supplied by the utility), the utilities can meet them during "off-peak" times by "filling" the solar energy storage systems with heat. Direct-gain passive systems, in this case, would not be able to store heat, though indirect-systems could have this capability. In short, the extra cost of storage may be worthwhile at some point, though it is not a critical factor today.

Solar energy passive design represents a potentially large and cost-effective resource. Unfortunately, many of the expenses related to solar energy passive design were excluded from the investment tax credit incentives of the National Energy Act. Thus, one of our cheapest and most readily available sources of solar energy has been neglected. Indeed, in the long term, it is conceivable that homes can be heated entirely by the internal heat gain provided by the waste heat of appliances, lights, occupants, and passive solar energy gain. Such homes would require very extensive insulation and air quality treatment features, but are plausible, nonetheless. For the short term, we should remember the close relationship of solar energy to energy conservation.

8.3.4. Utility-Owned Active Solar Energy Systems

Active solar energy systems, including heat pumps, are also conservation devices in the sense that they substitute for nonrenewable energy resources. The Annual Cycle Energy System (ACES),[17] in particular, is an electrically driven system which has been shown to provide building thermal energy requirements almost entirely from the sun. The system takes heat from the building in summer and stores it in large, very well insulated tanks. The heat is removed from the tanks in winter to heat the living space. Removal of heat from the water tanks creates ice, which is used for air conditioning in summer. If needed, extra heat is added to the tanks in winter by collecting sunshine on sunny days and melting the excess ice. Still, by the end of the heating season, enough ice is formed to carry the house through most of the cooling season.

The cost of such systems will possibly be more than most homeowners can afford, perhaps $10,000 to $20,000 per unit. But, in terms of energy supply capacity, the cost of an ACES system per unit of energy provided is comparable to that of a new electric generating plant. Utilities might well be persuaded to build new capacity in the form of ACES at the point of use instead of constructing central power plants, since external costs could be far less expensive. Such a scheme would be vastly preferable in situations such as in California where utilities would like to build power plants in Utah for Californians. If the cost is comparable and the service comparable, ACES, or other similar systems (passive systems would be cheaper) would certainly be more equitable in that the population of Utah would not suffer the damages of supplying California with power. A California utility, in fact, already engages in a similar practice by requiring that new residential customers must consent to have a solar water heater installed in their homes. The utility-owned ACES concept of a neighborhood-size solar system would simply be an extension of this arrangement. Utilities certainly should consider or be compelled to consider low- or no-interest loans to consumers for the installation of such systems. The lower cost of production would offset lost revenues.

8.3.5. Thermal Integrity of Existing Homes

Turnover in housing stock is not rapid. Only about one percent of existing homes are retired each year. We expect that three-fourths of all existing housing will still be in use in the year 2010. Because few homes have optimum conservation features, a retrofit program to improve the thermal efficiency of existing homes is essential.

Hirst and Jackson[8] have examined possible thermal integrity improvements in existing single family units in terms of increased capital investment. They first assumed that existing houses have conservation features equivalent to the 1970 HUD-MPS standards. This assumption was intended to underestimate the potential for conservation, since most homes are probably not so well insulated. At least 16 percent of American homes have no attic insulation whatsoever, including 14 percent of all homes in the northeast, and 9 percent in the north central region.[9] For such houses, the relationship between investment and conservation is far more favorable than for dwellings built to 1970 standards. That is not to say that the potential, even assuming 1970 standards, is not great. As Figure 8.2 indicates, retrofit expenditures on the order of $700 can reduce energy consumption for space conditioning in most existing houses by at least 40 percent. The payback on an investment of $700 in an uninsulated Kansas City home heated with electric resistance heat would be two years. The payback in a gas heated home would be, at most, seven years.

How does the homeowner choose among conservation options? If capable of making the capital investment, he or she should choose the option with the lowest total cost over the life of the investment. Here the consumer has to make a relatively sophisticated judgment. First, an assumption must be made regarding the future price of energy. Second, the cost of saving energy, including investment, loan, repair, and maintenance

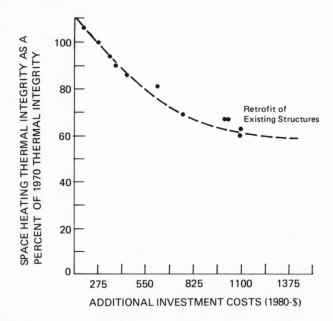

Figure 8.2. The cost of reducing residential space heating loads in existing homes.
Source: Reference 3.

costs, must be compared with the anticipated energy price. The consumer should invest in conservation up to the anticipated annualized energy price. Utilities are uniquely positioned to make financing available for these investments.

A classic study done by the Lawrence Berkeley Laboratory aptly ranks the value of four major conservation investments: automatic thermostats, ceiling insulation, wall insulation, and storm windows (see Figure 8.3). The study describes retrofit conditions in an uninsulated, gas-heated home in San Francisco. The value of conservation in a home located in a relatively mild climate using the cheapest energy source for heating obviously will be lower than in most cases in the U.S. But investment in every option except, perhaps, storm windows, is profitable. Natural gas, you will recall, costs about $3.00 per million BTU. The federal government, in fact, is subsidizing coal-gasification plants that produce gas costing at least $6.00 per million BTU. The cost of an automatic thermostat which would reduce temperatures to about 55 degrees during times when no one used the living space would equal the equivalent of $0.30 per million BTU saved—one-tenth the cost of natural gas, one-twentieth the cost of coal gas. Ceiling insulation (R-19, or six inches) would save energy at a cost of only $1.00 per million BTU. Wall insulation (R-11, or 3.5 inches) would cost almost as much as gas does now, but far less than it will in the near future. Storm windows cost about $4.50 per million BTU saved. Storm windows will be a

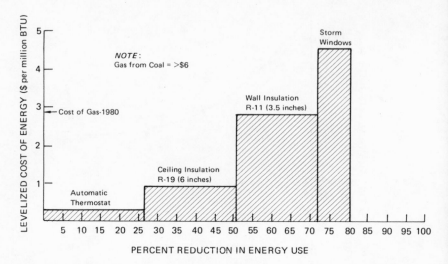

Figure 8.3. Residential energy conservation options—the cost of energy saved. Source: Reference 10.

bargain as natural gas increases in price under the rules of deregulation. For electrically heated homes, even those equipped with heat pumps, storm windows are a great bargain.[10]

Storm windows reduce both the conduction of heat through glass and the amount of air which infiltrates. Warm air that escapes and cold air which enters and must be heated are serious energy losses at windows, much more important than conductive losses. Caulking and weatherstripping, therefore, can be more effective than storm windows. Good construction is of paramount importance, not only for assuring that windows and doors fit tightly, but because sometimes as much as 40 percent of the heat leaked from houses escapes under poorly fitting soleplate (butts against which studs fit) and around electrical outlets.[5]

Table 8.5. Costs of Residential Retrofits with National Energy Act Incentives

Feature	Initial cost	Tax credit	Cost to consumer of energy saved ($ per million BTU)	
			without credit	with credit
Automatic thermostat	$ 125	$ 19	$0.30	$0.28
Attic insulation	450	68	1.25	1.05
Wall insulation	750	113	2.35	1.33
Storm windows	610	92	5.50	4.70
Total	$1935	$292	$1.65	$1.40

Awareness of the value of residential thermal energy conservation options and the incentives of the National Energy Act tax credits for conservation investments are prompting millions of housing retrofits. Table 8.5 shows the amounts of money a homeowner can save by investing in each of the aforementioned options before the end of 1985 when the tax credits expire. Credits may be taken on 15 percent of expenditures of up to $2000, for a maximum of a tax credit of $300. This is approximately the credit one would receive as an incentive to invest in the thermostat, attic insulation, wall insulation, and storm windows options described earlier. More or less investment would be justified depending on local climate and prices. If the National Energy Act goal of retrofitting 90 percent of the total 63 million home units by 1985 were met, then between one and two quadrillion BTU per year (the equivalent of one-half to one million barrels of oil per day) could be saved.

The availability of capital for this effort may be of some concern. To achieve the National Energy Act goal of retrofitting 64 million homes with equipment costing $625 per home (adjusted for inflation since 1978) would require the availability of $40 billion for loans and grants between now and the year 1985. This effort would reduce household unit thermal demands by an average of only 35 percent while we have shown that in many cases a 50 percent reduction is possible. The cost of achieving the larger reduction is about $1300 per house (omitting storm windows), for a total cost of about $83 billion. In comparison, the electric utilities expect by 1985 to be spending $55 billion per year to build new capacity and transmission and distribution systems,[11] money which generally would be tied up in loans amortized over 30 years. Home retrofit insulation loans would normally not extend more than ten years. The stronger conservation effort would have the effect of obviating the need for $30 billion worth of power plant capacity, assuming that all new demands were to be satisfied electrically. Thus, the issue of capital availability for conservation seems minor when cast in the perspective of capital requirements for new energy supply.

Tax credits for persons who do not make enough to pay taxes will do those persons little good, of course. The National Energy Conservation Policy Act of 1978* provides a grant program of $200 million per year in 1979 and 1980 for low income families to use for energy conservation. About 9 million households (single family units) have incomes below the

---

* Part of the National Energy Act. This Act also made it illegal for all utilities except those few with conservation subsidy programs in existence in 1978 to invest in energy conservation in their customers' buildings. This aberration of public policy was corrected, ironically, in the Synthetic Fuels Act of 1980.

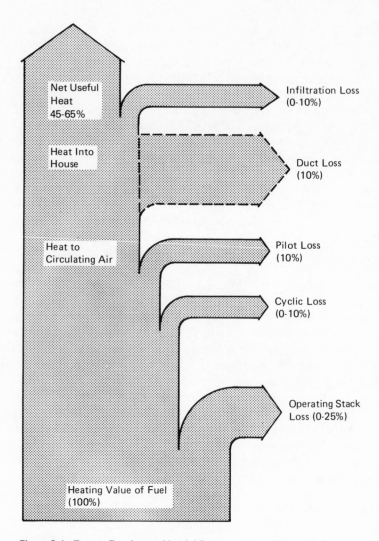

Figure 8.4. Energy flow in a residential furnace system. Source: Reference 1.

poverty level. The total cost of weatherizing 9 million homes would come to $6 to $12 billion, depending on the level of improvement. Viewed alternatively, the grant program would provide about $50 per low income house, or could fully retrofit 400,000 homes.* This level of assistance is woefully inadequate.

---

* A separate $25 million grant program will be made available for rural low income families through the Farmer's Home Administration.

Another major inadequacy of the National Energy Act is a failure to address the problem of weatherizing the large number of renter occupied homes. One-third, or 25 million, of all living units in the U.S. are rented.[9] It is unlikely that renters will make capital investments in dwellings they do not own, especially in multifamily houses. There is little incentive for landlords to invest in energy conservation when the occupant pays the energy bill. In fact, all the building owner is likely to get out of such an investment is higher property taxes, unless, of course, renters are willing to pay higher rents for thermally tighter buildings. An effective solution to this dilemma has been proposed by the city of Portland, Oregon, where no sale of a building will be permitted after a certain date unless it meets certain thermal standards. While this is strong medicine, renters deserve strong protection from inefficient dwellings, and the energy dilemma itself requires bold action.

8.3.6. Upgrading Gas Furnaces

Natural gas furnaces usually are described as at least 65 percent efficient. In reality, most existing units probably deliver half the chemical energy in gas to the living space as heat. As Figure 8.4 illustrates, large losses are incurred up the smoke stack, through operation of the pilot light, with duct losses, and infiltration. Such losses amount to 25, 10, 0–40, and 0–10 percent, respectively. Improvements in gas furnaces such as using heat recovery in the stack, replacing the pilot light with an electric ignition system, and insulating heat ducts could reduce gas use per unit by one third.[13]

8.3.7. The Thermal Integrity of Commercial Buildings

Commercial buildings use about 12 percent of the country's energy, and represent one of the fastest growing categories of energy demand. Four building types, retail–wholesale, office, health care, and educational buildings, consume 71 percent of the energy used in this category (24, 16, 12, and 19 • percent, respectively). Heating accounts for 42 percent of demand, with lighting and cooling requiring 23 and 21 percent, respectively.[14] Both theoretical and empirical results show that energy use in new buildings of this type could be reduced by 40–60 percent at no net additional construction cost.[12] Since many of these buildings are publicly owned, notably school buildings, the opportunity for savings should be recognized.

A federal office building in Manchester, New Hampshire, exemplifies this opportunity. A seven story building was commissioned in 1972 by the U.S. General Services Administration to serve as a model for conservation. Design features included no north facing windows, reduced overall window

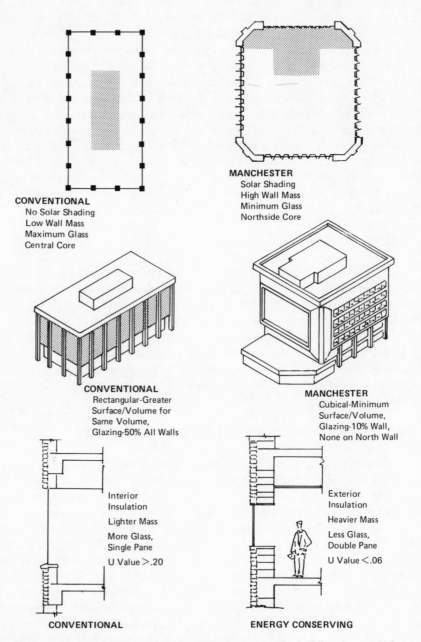

CONVENTIONAL
No Solar Shading
Low Wall Mass
Maximum Glass
Central Core

MANCHESTER
Solar Shading
High Wall Mass
Minimum Glass
Northside Core

CONVENTIONAL
Rectangular-Greater
Surface/Volume for
Same Volume,
Glazing-50% All Walls

MANCHESTER
Cubical-Minimum
Surface/Volume,
Glazing-10% Wall,
None on North Wall

Interior
Insulation

Lighter Mass

More Glass,
Single Pane

U Value > .20

Exterior
Insulation

Heavier Mass

Less Glass,
Double Pane

U Value < .06

CONVENTIONAL   **ENERGY CONSERVING**

Figure 8.5.   Energy conserving vs. conventional design in an office building. Source: Reference 12.

Table 8.6. An Illustration of Commercial Sector Thermal Energy Conservation Options

Feature[a]	Initial cost	Energy saved per year (million BTU)	Cost of energy saved (1980 dollars per million BTU)
Central automatic thermostat control	$124,900	15,500	$0.90[b]
Use reject chiller heat for reheat in summer	14,200	2,300	0.70
Exhaust air controls	151,700	11,700	1.00
Return air system with economizer	34,700	2,930	1.00
Exhaust air heat recovery	111,230	4,950	2.60
Insulate/reinsulate steam piping	14,800	960	1.90
Automatic light switches to increase use of natural light	4,500	60	8.60[b]
Additional roof insulation (during reroofing)	41,360	740	6.55

Source: Reference 12.
[a] Selected features described in Reference 12.   [b] Should be compared with cost of electricity.

area, increased insulation and thermal mass, a heating and ventilation system which allows exchange of air among over and underheated areas, and others (see Figure 8.5). No additional construction cost was incurred, while savings as high as 20 percent over other new local office buildings were achieved.[12]

Similarly, retrofit options for other commercial, educational, or institutional buildings offer great potential. An example is the retrofitting of a chemistry building in Brookhaven National Laboratory. Some space heating features, their initial cost, and the cost of energy saved are shown in Table 8.6. Many cost less than natural gas, which costs commercial users an average of $2.20 per million BTU. Thermostats, heat recovery systems, insulation, automatic switches, all are cost effective for use in public buildings.

The incentives offered in the National Energy Act to facilitate such improvements included:

o   A $900 million *grants* program to schools and hospitals awarded during 1979–1981.
o   A $65 million program for performing audits on public buildings (2 years).
o   Design standards for new federal buildings.[4]

No tax incentives have been provided for energy conservation investments in businesses. The rationale behind this decision might have been that businesses, unlike homeowners, are out to make a profit and therefore act "rationally" and do not need conservation investment incentives. In any case, it seems prudent that the federal government require construction standards stricter than the ones used in the Manchester Building, which are similar to the American Society of Heating, Refrigeration, and Air Conditioning Engineers (ASHRAE) 90-75 voluntary standards, and *far* less than optimum. Strict performance standards should also be applied. This is not to say that thermostats should be set at uncomfortable levels; there is no need for discomfort if rational investments are made.

Because the commercial buildings sector is so large and represents part of the fastest growing component of the American economy, the service sector, and conservation opportunities within it, demands greater attention.

8.4. Water Heating

Water heating is the second largest category of energy demand in the buildings sector. Requiring 11 percent of the energy used in this sector, and four percent of all energy consumed in the nation, water heating accounts for nearly two quads of energy use annually. The energy thus consumed excludes the primary energy required to generate electricity for electric water heaters which represent approximately 50 percent of all U.S. water heaters. Opportunities for saving energy in domestic water heating* include (1) improving the thermal efficiency of gas and electric water heating, and (2) using solar water heaters.

8.4.1. The Thermal Efficiency of Residential Water Heaters

Conventional electric water heaters can be upgraded to reduce energy losses by adding insulation to either the storage tank or distribution system. Figure 8.6 depicts the relationship between increased initial cost of an electric water heater and its energy consumption. The figure shows several means of reducing a water heater's annual energy consumption from about 23 million BTU per year to less than 20. The less efficient water heater costs about $220 new. Adding 7.6 centimeters (about 3 inches) of fiberglass jacket insulation, for example, will reduce the energy demand for water

---

* We concentrate on domestic water heating because water heating in commercial buildings accounts for only 2 percent of that sector's demand.

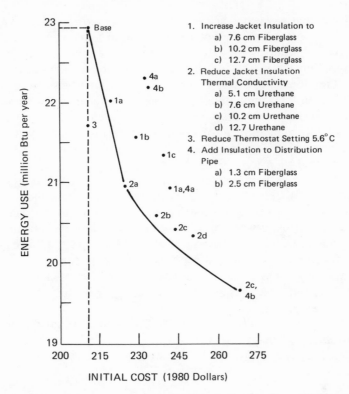

Figure 8.6. Conservation options for electric water heaters. Source: Reference 15.

heating by about one million BTU per year, but will raise the initial cost of the water heater by about $10. The tradeoff here between extra first cost and less energy cost saves energy at less than $1.10 per million BTU. The use of a better insulator like urethane foam, however, will reduce energy demand in the electric water heater by almost two million BTU per year for only a few dollars more; the cost of energy saved would equal only $0.80 per million BTU. Adding insulation to the water piping throughout the house interior would not be cost-effective, however, unless electricity were priced at $14 per million BTU. Note that a simple reduction in thermostat setting would cost nothing and would save more than one million BTU per year. A setback of 5.6 °C (10 °F) would not normally cause any discomfort. Many water heaters are set to heat water to 65 °C (150 °F), although human skin cannot without pain tolerate water at temperatures greater than 46 °C (115 °F). Dishwashers, however, do need to operate with water at slightly higher temperatures. New dishwashers should have built-in capability of heating water to the necessary level.

Conventional gas water heaters can be improved in similar ways. In addition, heat losses due to pilot light operation and to the flue can be reduced. Simply reducing the rate of burning of the pilot light can (safely) save two million BTU per year, savings comparable to the reduced thermostat setting. Reducing excess air in the combustion of gas will save more than one million BTU annually. These operational changes alone can save five million BTU per year, a savings on an individual's gas bill of $15 per year. An additional investment of $10 in urethane insulation would add an annual savings of $10 per year. The payback period is, thus, one year, even at current natural gas prices.[15,16]

The electric heat pump water heater is a device which operates in the same manner as any heat pump, except that it heats water rather than air. Heat pump water heaters were first marketed in the 1950s but were withdrawn because demand was low and because of technical problems. Now they are being reintroduced at a cost of about $700. Since they are at least twice as efficient as conventional water heaters, these new systems should provide annual savings of $100 and would thus have a simple payback period of five years, less in high electricity energy cost areas such as New York.[15]

An Annual Cycle Energy System (ACES) would reduce the need for externally supplied energy for water heating by 70 percent. Unlike other options we described, however, ACES is not yet cost effective for single-family residences.[17]

8.4.2. Solar Water Heaters

Residential solar water heaters are initially expensive. They cost about $2000 each and deliver hot water for a cost of $5 to $15 per million BTU, depending on location and building type. In most parts of the U.S., solar water heating is already competitive with new electrical energy capacity for water heating. With the 40 percent tax credit made available on the first $10,000 of investment in solar equipment due to the National Energy Act,[4] solar water heating is a bargain in many places. Problems such as fluid leaks, dirty collectors, snow, and so forth have been experienced by the growing number of owners, but none of these problems seems overwhelming or insoluble.[18]

The most productive efforts to improve the energy efficiency of water heaters, to summarize briefly, are to insulate the storage tanks, reduce pilot and flue losses on gas systems, and to substitute heat pump water heaters and solar water heaters for conventional systems, whenever possible. The National Energy Act has fostered these improvements in new systems; however, the job of retrofitting with insulation the tens of millions of hot water tanks in the nation is a job neglected. The individual's stakes are rather small, a few dollars per year, but the aggregate savings to any utility system, gas or electric, could be very great. Policy tools for encouraging utilities to take on this job need to be applied.

8.5. Air Conditioning

Air conditioning, the third major energy end use in buildings, requires more than one quadrillion BTU per year in the U.S. Energy policymakers, which certainly include every home builder and owner, should recognize the overriding importance that insulation and thermal integrity have for space cooling and heating. You will recall from Section 8.3.1. how improved residential insulation reduced both the need for energy for air conditioning and the need for air conditioning equipment in new homes. The benefit-cost evaluation each decision-maker conducts regarding household thermal integrity improvements should include the increased comfort and reduced costs of air conditioning as well as heating.

Air conditioning appliances also offer room for improvement. A measure used for the performance of air conditioning units is the ratio of cooling capacity to the power requirements (BTU per hour divided by watt-hours), or the Energy Efficiency Ratio (EER). The average EER of window units on the market today is about 6, although it ranges from 5 to 12. The market price per unit apparently is not related to efficiency. California has set a minimum performance standard of 8 BTU/watt-hour for new sales. The U.S. government should mandate a move to higher efficiency units with appliance efficiency standards as authorized in the National Energy Act. The effect could thus be as great as 50 percent energy savings per unit.[1]

8.6. Lighting

Chief Bromden, a character in *One Flew Over the Cuckoo's Nest*,[19] thought that white fluorescent lights were refrigeration coils designed to freeze the soul. Maybe not, but the intensity at which they are used in many places does cause eye strain. Underillumination, on the other hand, seems to have no ill effects, contrary to still another popular myth. Lighting energy requirements total almost one quad annually. Significantly, it accounts for about 20% of all electrical demand. Thus, important savings can be had by reducing unnecessary lighting, performing lighting tasks with less primary energy, and also decreasing building heating loads due to the waste heat of light bulbs.

It is true, however, that fluorescent light tubes provide three times more light per unit of energy consumed than incandescent bulbs. So, here is one case where aesthetics appears to clash with energy conservation. The way fluorescent lighting is used, however, may counter its inherently greater efficiency. The use of saturation lighting—the illumination of every corner of every room as if one had to thread needles everywhere—uses energy unnecessarily and adds heat to the air conditioning load in summer. In winter, space heat can be supplied by much more efficient systems than waste heat from lights. Indeed, the cores of many buildings must be cooled in winter partly because of too much lighting.

The saturation approach to lighting can waste up to 50 percent of the energy lighting requires and in summer adds 0.4 watts to the air conditioning load for every watt of lighting.[12] The slow growth in peak electrical demand experienced recently may be due in no small measure to the reduction of intense lighting levels focused in commercial buildings.[20] Peak electrical demand in most utility systems* occurs with summer air conditioning loads. Clever design, which should be prompted with building codes, would incorporate task-oriented lighting, the use of sunlight, and switching devices to turn off lights during times when the space they illuminate is unoccupied, or when sunlight makes them unnecessary. The cost of energy saved from such devices ranges from $0.15 to $0.85 per million BTU in new buildings, not counting the air conditioning savings. Retrofitting existing buildings with light switch timers would save energy at a cost of no more than a dollar per million BTU.†[12]

The advent of new lighting technologies may illuminate more efficient energy paths. The new LITEK bulb is a fluorescent light which is three times as efficient as the incandescent bulb, yet shines with a gentler, broader spectrum, more natural light, and works in incandescent sockets. Its cost is about $7.50 per bulb. It has a payback period, assuming several hours of use per day, of about two years, and fortunately has a lifetime of ten years.

8.7. Cooking

The process of preparing food requires about one-half quadrillion BTU each year. Most of this energy is consumed in residences, although restaurants and industrial food preparation operations do require large amounts of energy. For large users of energy for preparing food, two systems offer the opportunity for cost reductions.

---

* TVA, the nation's largest utility, is an exception. The nation's largest utility has a *winter* peak due to its successful promotion of electric heating.

† The difference in cost between new and retrofit buildings is the difference in time over which the retrofit must be discounted. A ten-year life is assumed for existing buildings, compared to 30 years for new buildings.

Microwave ovens are already a widely known and used technology. They operate on the principle of microwave transmission through foods, which has the effect of causing water molecules to oscillate. This oscillation generates heat which cooks the food evenly, swiftly, and with far less energy than otherwise. Microwave ovens reduce cooking energy consumption by 20 to 40 percent. Unfortunately, they will not brown the exteriors of foods.

Certain food preparation plants and large hotels have taken advantage of district heating systems for low cost energy for cooking. District heating is a concept used widely in Europe, in which steam is generated in a central plant and distributed for space heating, industrial uses, and other purposes. Hot water can be transported economically for up to 100 miles.

The Brock Candy Company in Chicago has contracted with the city of Chicago to purchase steam for candy cooking from Chicago's municipal incineration plant for $3 per million BTU, much cheaper than Brock could buy oil and steam generation equipment. The Hyatt Regency Hotel in downtown Nashville cooks with steam delivered by pipeline from the Thermal Transfer Corporation, which operates a garbage incinerator for the city of Nashville. Many college campuses have district heating systems and use some of the heat for cooking. Although American cities may be too sprawling to use district systems economically for heating, cooking, and other purposes, industries and large institutions can certainly make use of them at considerable savings.

8.8. Refrigeration

Refrigeration equipment, which uses about one-third of a quad annually in the U.S., can be considerably improved for energy efficiency. Figure 8.7 indicates that household refrigerators could be twice as efficient without sacrificing any service or convenience, and without any extra total cost. The figure lists eight options which add efficiency without loss of service. These include increasing insulation, removing the fan from the cooled area, adding an anti-sweat switch, improving compressor efficiency, and increasing the surface areas of the evaporator and condenser. A combination of all these improvements would add about $100 to the purchase price of the refrigerator. The energy savings, however, would total more than 8 million BTU per year. Since electricity costs an average of $10 per million BTU, the improved, though more costly, refrigerator would pay back the extra expense in 16 months (12 months in Boston). *In fact, the new refrigerator would pay for itself totally through energy savings over seven to eight years.*

The "frost-free" feature stands out as an expensive convenience. Note that adopting option 5, elimination of the frost-free and forced air features,

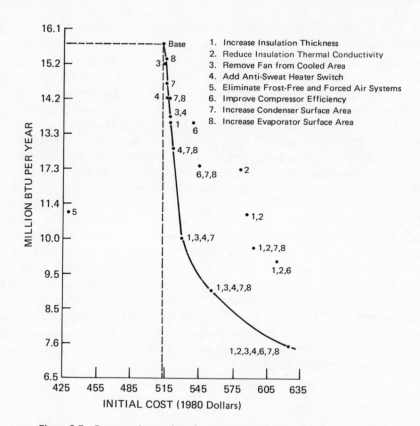

Figure 8.7. Conservation options for the refrigerator. Source: Reference 21.

would save nearly $50 worth of electricity (5 million BTU worth $10 per million BTU) each year plus $80 on the purchase price.

Refrigerators, then, appear to be good targets for improved consumer awareness or federal appliance efficiency targets. Because refrigerators are not retired particularly rapidly, the energy savings impact will not be rapid. In the long term, however, the savings can be significant, the equivalent of about 25 million barrels of oil per year, more than three day's total of oil imports. This savings, we hasten to point out, assumes that the frost-free feature is not eliminated.[21]

8.9. Summing up the Savings

Residential and commercial demand for the amenities that energy helps provide will continue to grow. The U.S. population may increase from the current 214 million people to 250 million or more by the year 2010. The number of households will likely increase from 72 million to about 110 million, with single family dwellings, which represent 47 million of the present total number of housing units, growing to 67 million by that year. Commercial floorspace will grow to 55–78 billion square feet, though demand for educational space will diminish with the increasing age of the population.

The real income of individuals will probably increase by a third over the next three decades. Each of these trends points to higher demand for energy in buildings. But we have illustrated the economic feasibility of reducing energy use in new homes by 40 percent over what is called for by present standards. Existing homes could be retrofit to reduce heating and cooling requirements by 50 percent. Water heaters and air conditioners energy demand could be cut in half. Other appliances, many of which are approaching market saturation, can be substantially improved. We cannot realistically expect that all of this potential increase in energy productivity will be realized; neither could we simplistically add up the savings. Modeling efforts such as those of CONAES,[1] and recent studies by the Office of Technology Assessment (OTA),[5] by the Council on Environmental Quality,[22] and by the Mellon Institute's Energy Productivity Center[23] are required to cross safely these minefields of assumptions.

The CONAES results reflected a certain pessimism about our ability to adjust to the new energy realities, even where failing to adjust would be to our detriment. For new homes in 2010, the CONAES "B Scenario" assumed a reduction of only 33 percent over 1975 new home performance. Retrofit houses would be 35 percent tighter thermally than they were in 1975. These are enormous but readily achievable improvements. CONAES concludes also that electric water heaters would be improved by 15 percent, that gas-fired furnaces would be made 25 percent more effective, that refrigerators, which can now be built 40 percent more efficiently, would be improved by 32 percent, and that air conditioning efficiency would be improved by only 30 percent. In comparison with what is now economically feasible, none of CONAES' assumptions appears to be overly optimistic. "Scenario B" is quite cautious. "Scenario A," a future in which real energy prices would quadruple by 2010, was more aggressive, of course, but not overly bold. Table 8.7 compares these two estimates of future energy economy with the maximum economically feasible today for several uses of energy.

What these comparisons mean is that total buildings sector energy demand will not be as high as 34.3 quads, as USBM-1975 predicted. Perhaps demand will be "only" 9.5 to 12.6 quads in the year 2010 (compared with 16.8 in 1975) as CONAES "A" and "B" results indicate. OTA reported similar results with similar energy price assumptions. Perhaps,

Table 8.7. Energy Intensities: The Technical Potential and the
CONAES Projections (Indexed: 1975 = 1)

Item	Year 1975	Year 2000 "Scenario A"	Year 2000 "Scenario B"	Present technical and economic potential
Thermal integrity				
New homes (single family)	1.0	.51	.67	.40[a]
Retrofit homes (single family)	1.0	.60	.65	.50[b]
Heating appliances				
Electric	1.0	.57	.66	.2[c]
Gas	1.0	.72	.73	.69[d]
Water heaters				
Electric	1.0	.85	.85	.50[e]
Gas	1.0	.73	.75	.50[f]
Air conditioning	1.0	.66	.70	.50[g]
Lighting	1.0	.65	.70	.35
Refrigerators	1.0	.58	.68	.43[h]

Source: Reference 1.

[a] Assumes attic insulation of R-49, wall insulation of R-23, floor insulation of R-19, triple-paned windows, and storm doors. Additional capital investment totals $2195 when house is built.

[b] Assumes $700 retrofit investment in a typical 1500 square foot single family dwelling.

[c] Assumes installation of an annual cycle energy system, and that these will be commercially available in the 1980s.

[d] Assumes improvements in conventional gas furnace systems such as heat recovery and replacement of pilot light with electric ignition. See Chapter Eight.

[e] Electric heat pump water heaters.

[f] Gas heat pump water heaters.

[g] Assumes an Energy Efficiency Ratio (EER) of 12, compared to the existing availability of air conditioners of EERs of 5–12.

[h] Assumes an extra $130 additional capital investment per new refrigerator, and that the frost-free option *is* included.

with volition, but without discomfort or loss of amenities, we could drive the energy consumption of buildings even lower. The CONAES "A*" scenario, in which nonmarket steps to curb consumption such as Portland, Oregon, took in mandating that no buildings which failed to meet certain standards could be sold after a certain date, resulted in an estimate of energy demand in the buildings sector in 2010 of only 6.4 quads.

8.10. Meeting Demand

If it is imperative that the buildings sector energy demand for the year 2010 total no more than 10–13 quads from all sources, then we argue that there are any number of ways to meet that goal. Gas now accounts for 8.5 quads to the buildings sector; oil and electricity provide 4.1 and 3.7 quads, respectively. Switching the buildings sector to electricity would be a good strategy to conserve oil, particularly if the new source of electricity were cogeneration plants. Natural gas, in fact, would be more efficiently used in industrial cogeneration than in residential furnaces. However, if deregulation of natural gas makes new gas supplies available, these will almost certainly be cheaper than electrical energy. Similarly, synthetic gas from coal would be cheaper than electricity from coal.

8.11. Research and Development

The options for residential and commercial energy conservation described in this chapter are all either commercially available or nearly so. Yet, even greater conservation is possible by advancing the commercialization of certain technologies:

o   Better heat pumps
o   Improved insulation
o   Lower cost solar equipment
o   Neighborhood-size solar systems
o   Fuel cells
o   Microprocessor controls for heating, lighting, and other appliances

Most of these technologies could conserve energy at a cost far lower than advanced energy supply systems. While we believe that it is the U.S. government's role to undergird basic research which would be either too expensive or else would pay dividends too far into the future for the private sector to support, the government should also help set the requisite climate (via market incentives, energy prices, federal purchasing practices) for the private sector to develop and commercialize such equipment. There are, in addition, major opportunities for research related to more efficient energy use.[24] Such research merits vigorous and sustained support.

8.12. Summary

Population, income, and new energy-consuming devices, the factors that help drive up energy demand in the buildings sector, should continue to grow until some time in the early twentieth century, but at much slower than historical rates. Saturation in certain types of demand (e.g., refrigerators), together with the impetus of higher prices and the use of solar energy will cause energy growth in the sector to be sharply lower, perhaps even negative. Increased electrification in this sector is likely and desirable insofar as petroleum can be economically conserved. Work on new conservation technologies could offer greater conservation opportunities than we have implied.

There is much slack in our energy system. There is far more than an imperfect market and even the best communications and educational system can take up. To tighten our use of energy in buildings will leave us still with the benefits of energy services but will reduce the threats of a diminished environment, of unhealthy contaminants, and of war. The object of energy policy, we believe, should not be to provide energy for energy's sake, but to provide the needs of people at costs that can be afforded. Energy conservation is and will remain the cheapest means of providing additional energy needs in the buildings sector for many years to come.

(Cars) are freedom, style, sex, power, motion, color. . .everything is right there.

—Tom Wolfe
The Kandy-Kolored Tangerine-Flake Streamline Baby (1965)

I would supply gasoline only to those motorists who could give a reason for wanting to drive somewhere. That would eliminate much of this driving about from one point to another by persons interested solely in atrophy during mobility.

—E.B. White
Quo Vadimus, Or The Case for the Bicycle (1927)

Chapter Nine

The Crisis and the Car

9.1. Perpetual Motion

Tom Wolfe called the American automobile a baroque extension of the ego. He called it freedom, style, motion, sex, power. . .everything. For many Americans, it is a virtual necessity. And whatever else it is, it is the nemesis not just of the conservation of oil, but of air, land, and lives. The automobile is the major source of four of the six most serious air pollutants. It has made possible or perhaps necessary the paving of five percent of America. It costs 50,000 human lives and millions of injuries each year. One thing to keep in mind for energy policy is that only the energy part of these problems is new. If we were unwilling to curb our cars to save hundreds of thousands of lives, among other things, will we be more willing to do so to save BTU?

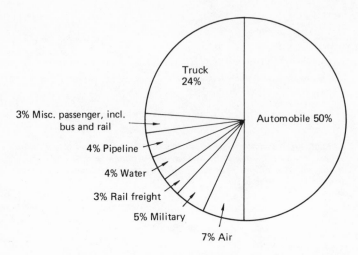

Figure 9.1.   Energy use for transportation in the U.S., 1978. Source: Reference 1.

9.2. Conservation Challenged

Almost half of all the oil used in the U.S. is used in transportation. Private motor cars are the main focus of this chapter on conserving transportation energy because in many respects the automobile is the energy crisis (see Figure 9.1). We import five million barrels of oil per day and use six million barrels of gasoline per day in cars. One out of every nine barrels of oil used in the world is burned in American auto engines. There are both social and technical fixes to this problem. Curtailment of vehicle use is, of course, an option, the only quick option we have in the event of an oil embargo. Fuel switching, too, is an alternative, and some grain alcohol from corn has replaced a token amount of imported oil. But grain alcohol is certainly not a long-term solution; neither are synthetic fuels from coal or oil shale a short-term (i.e., within this decade) solution. Conservation is thus challenged as much in the transportation sector as anywhere in the economy to make good on its claims. Can we substitute capital, design, and new techniques for gasoline in a meaningful way? The answer is unambiguously "yes."

This chapter focuses on the technical options for improving the car: weight reduction, engine efficiency improvements, new transmissions, and so forth. Attention is also paid to changes in automobile use, for emergency situations may have us calling off all bets. But here, as always, the issue is providing the amenity, transportation in this case, with as much personal freedom and as little cost as possible. Certain technical improvements could be greatly accelerated, however, under stressed conditions.

This chapter also focuses on the use of the transport truck, its inefficiency and alternatives. Trucking consumes 24 percent of the energy used in the transport sector.

Mass transit is evaluated both in terms of its theoretically high energy efficiency and in the light of its value for improving our quality of life. It is evaluated also in terms of its poor efficiency and high cost in actual practice.

Over all discussions of energy there hangs the specter of disruption and war. Many assertions have been made that certain opportunities such as synthetic fuels production and/or vastly more efficient cars must be undertaken for national security reasons, regardless of cost. This "damn the torpedoes" mentality ignores the fact that even in war there must be some notion of cost-effectiveness, of how much can be accomplished in a given time, and of certain investments that may have positive or negative benefits beyond immediate concerns. A crisis probably will do nothing to alter the rank order or value of energy options, although the saleability of certain options may change. The conservation options we discuss here would, if anything, become more valuable in the event of a major oil crisis.

9.3. The Car in Crisis

The personal car, despite criticism for its role in the urban crisis, the environmental crisis, and now the oil crisis, has changed little in basic design over its history. Basic changes that have come about were made for style and convenience. No one today has to hand crank a car, or even move a foot or hand a few inches to change gears. Power accessories take virtually all the effort out of mobility. Now there is serious discussion of the need for a cleaner, more efficient engine, one that could burn a variety of fuels, of the need for a more efficient transmission, and of the need for bringing the electronics and materials revolutions to the automobile. There is, in other words, serious discussion of the need to reinvent the car.

The motivation to switch from the spark-ignition, internal combustion engine, invented by Nikolaus Otto in 1891, originates in the desire to reduce emissions of air pollutants while maintaining car performance and fuel economy. It seems clear that certain engines, especially the Stirling engine, could achieve both better fuel economy and lower emissions, though it is less clear that the alternatives would represent improvements great enough to warrant the effort and expense of making the switch.

9.3.1. Automobile Efficiency Targets

In today's cars, about 12 percent of the energy in the gasoline reaches the rear wheels. The rest is dissipated in the exhaust, in engine or drive train friction, or in other frictional losses. Much of what reaches the wheels is lost in tire flexion and overcoming aerodynamic drag. Energy losses which have been measured at specific points include[2]:

o   exhaust, 33%
o   cylinder cooling, 29%
o   air pumping, 6%
o   piston ring-cylinder friction, 3%
o   accessories, 2.5%
o   transmission, 1.5%
o   axle, 1.5%
o   braking, 3.5%
o   coast and idle, 4%
o   tires/rolling resistance, 6%
o   aerodynamic drag, 6%

For a car of a given weight and performance level, at a steady mode of operation, these losses can be decreased in ways illustrated in Table 9.1A. Listed are specific improvements which may be achieved in the areas of

Table 9.1A. Possible Automotive Fuel Efficiency Improvements[a]

Option number	Specific improvement	Reduction in automobile energy consumption (percent)
0	Weight reduction (from 4200 pounds to 2400 pounds)	40
1	Aerodynamic drag reduction	10
2	Piston ring-cylinder friction reduction (of 33 percent)	3
3	Adiabatic (i.e., not cooled) diesel engine	40
4	Continuously variable transmission	20
5	Continuously variable transmission with regenerative braking and energy storage	35
6	Improved engine and drive lubricants	5
7	Low drag tires (25 percent reduction in rolling resistance)	13
8	Traffic management for reduced idling	20
9	Stirling engine	20–50
10	Improved Otto cycle engine	25
11	Valve resizing	13
12	Xylan engine coating (50 percent friction reduction	15
13	Idle off systems	12

[a] Options are not additive; see Table 9.1B.

Table 9.1B. Automotive Fuel Savings with Combinations of Possible
Improvements

Combination of improvements (see Table 9.1A. for definition)	Percent reduction in automobile energy consumption
1 + 7	20
9 + 5	54
9 + 5 + 1 + 7	67
3 + 1 + 7	55
9 + 1 + 7 + 12	53
10 + 1 + 7 + 12	50
10 + 1 + 7 + 8	55
10 + 4 + 1 + 7	52
2 + 1 + 7 + 11 + 12	39
1 + 2 + 7 + 6 + 11 + 12	41
1 + 7 + 9 + 11 + 12 + 13	43
11 + 12 + 1 + 7	37

Source: Adapted from Reference 2.

aerodynamics, tribology (friction), rolling resistance, traffic management, combustion research and development, and advanced materials application. Table 9.1B illustrates the potential savings which could accrue from the combination of certain of these options. Note that improved aerodynamics and tires (options 1 and 7) alone could reduce automobile consumption by 20 percent, and that adding improved Otto cycle engine efficiency and a highly efficient transmission would save 52 percent of the energy used in a car. These examples (options 1, 4, 7, and 10) are presently available technology but cannot be expected to be generally available to new car buyers until 1990. Yet it is possible that perhaps some of the potential of these examples will be used in complying with the 1985 federal gas mileage requirements currently being met with weight reductions and engine downsizing. Any given option to penetrate most of the automobile fleet would require seven to ten years lead time simply because about 10 million of the more than 100 million cars in the U.S. are replaced each year.[3] Thus, it is imperative that standards be set as soon as possible for improved fuel efficiency. If Detroit must have five years or more to plan, design, and retool to produce cars of a different type, and if it has already made huge commitments, as it claims, to meet the 1985 standards, then we must act now to set standards for 1988–1990. We do believe that standards are necessary, for the total expense of owning and operating an automobile is such that it makes little difference in cost to the consumer whether he or she buys a car that provides 17 miles per gallon or 37. We turn to the economics of this issue in Section 9.3.2.

Other potential improvements are abundant. Twenty-nine percent of all gasoline energy is lost in cylinder cooling. An adiabatic diesel engine (one that was not water cooled) would cut these losses considerably. A diesel engine is inherently more efficient because it can operate at a higher compression ratio, but we will return to these potentialities in a later section on engines.

Other options include reducing cylinder friction and air pumping losses. The Otto engine is a pump that compresses air to be ignited with fuel. For reasons we will discuss later, these losses are worse when the engine operates at partial load, as it does most of the time. It has been estimated that automobile engines run at less than 10 percent of full power more than 40 percent of the time.[4] New transmissions and electronic engine control can effect huge savings by more appropriately matching the engine and power requirements.

Electronic control of the engine also promises great savings in other ways. With four percent of all automotive energy lost in idling, usually in traffic in cities where engine-fuel emissions cause great harm, electronic control can be used to shut down certain cylinders, or all of them for that matter. Precise control could allow instant restarting.

The list of potential improvements in automobile fuel economy is long. The cost and technical feasibility of certain of these options should suffice to illustrate the potential.

9.3.2. The Cost of Automotive Energy Conservation

It makes little difference in terms of total automobile operating costs whether a consumer purchases a car that produces 20 miles per gallon or 40.[5] Although gasoline for the 20 mpg car would cost (assuming $1.00 per gallon) five cents per mile compared to only two and one-half cents per mile for the 40 mpg car, the difference is small when measured against nonfuel costs which total more than twenty cents per mile. Figure 9.2 portrays this condition graphically. Note how the top curve on the graph, which shows the relationship between total operating costs and fuel economy, is relatively flat, even declining slightly, between 20 mpg to 40 mpg. Whether gasoline costs $1.00 per gallon or $1.85 per gallon, the car with the least cost is the one that obtains 37 mpg. *But the difference in total cost is less than 10 percent over a very broad range.* This fact, coupled with the urgent need to reduce oil imports, is a single, powerful justification for minimum fuel economy standards. Since consumers are not likely to see the effect of price unless the price of gasoline skyrockets, fuel efficiency standards far higher than those set for 1985 are called for. A 37 mpg standard by 1990 should be a minimum goal; 50 mpg is not out of the question.

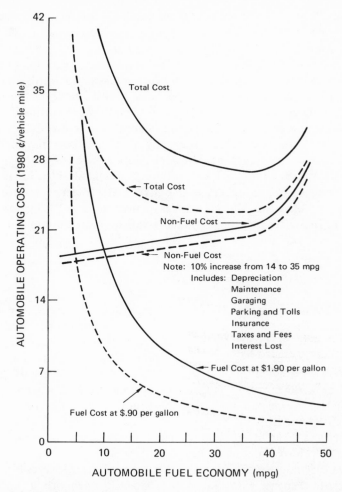

Figure 9.2. Total automobile operating costs vs. fuel economy. Source: Reference 5.

A large part of the question revolves around the impact on the ailing automakers of the cost required for reorganizing to make efficient cars. Chrysler and Ford have severe cash flow problems because they persisted in trying to sell large cars for which the market had largely disappeared. It may be true that these manufacturers would be worse off had they not been forced toward more efficient designs. The case of the automobile fuel economy standards is a rare example of how the government, which took a broader, longer perspective, improved public welfare through market intervention. A similar response was required to effect emissions controls. The question, of course, arises, "How far should the government go in sheltering the automobile industry from the shocks of investing in new, fuel efficient cars?'' It is our own belief that manufacturers, as a rule, should not be rewarded for making mistakes, though it should be pointed out that federal gasoline price controls contributed greatly to the decision to produce and purchase inefficient cars. Stringent fuel economy standards, and we are thinking of standards above 40 mpg, may be justifiable in the national interest, in the broader economic sense, though they are not justifiable in a narrow economic or market sense to automakers and individual consumers. In the latter case, capital subsidies by the U.S. government for the automobile industry might be defensible, just as it has been argued that such subsidies of alternate energy sources are justifiable. In any cost-benefit comparison, however, automotive fuel efficiency improvements come out far ahead of alternative energy investments. Perhaps more importantly, fuel efficiency improvements will be more certain and can be achieved more rapidly.

9.3.2.1. Weight Reduction

Fuel use in cars is almost directly proportional to weight. A good rule of thumb is that a car will use one gallon of gasoline over its lifetime for every pound of weight. If all 100 million plus American cars weighed an average of 1000 pounds less, four quadrillion BTU could be saved each year. Stated another way, a car weighing 2000 pounds will use half as much gasoline as a 4000 pound car, all other things equal. Clearly, weight reduction is the principal option for fuel economy.

At this point, some will object that weight reduction in cars conflicts with other goals like safety, comfort, and pleasure, amenities that may be more highly valued than fuel economy.

The objection that smaller cars are less safe than larger cars is probably the easiest to dispense with. Safety is as much, if not more, a function of maneuverability, design, and operation as it is of weight. Lighter cars are less likely to be involved in a crash. Small cars can be equipped with specially designed bumpers, I-Beams in the doors, collapsible steering wheels, reinforced roofs, etc., in order to protect passengers better than do most large cars today. It is true that these features carry weight penalties, but these can be offset elsewhere.

Options for reducing the weight of cars without reducing passenger or baggage space include materials substitution, engine down-sizing, and front-wheel drive. Materials substitution, the use of lighter but more expensive metals such as aluminum and magnesium to replace steel, has been estimated to be capable of reducing fuel consumption in American cars by eight percent at a cost of about $50 per car.[6] This savings will cost the first owner, who usually owns a car for three years, the equivalent of $0.50 per gallon of gasoline. The savings are free for the subsequent years.

Matching the engine and power requirements allows the use of lighter engines. Lost power, due to smaller engines, can be replaced by turbocharging.

Advanced Otto cycle engines could be reduced in weight by the use of ceramics and more compact configurations. Savings of perhaps 350 pounds could be realized without sacrifice of power.[7]

Front-wheel drive obviates the need for a crankshaft running the length of the car from the engine to the back wheels. Weight can thus be saved, while at the same time the passenger compartment can be made roomier since the hump down its center can be eliminated. General Motors recently installed manufacturing capability for a front-wheel drive car (the X-Car) but claims the cost of producing this new vehicle totals $2.6 billion.[8] Not all of this cost can be credited against fuel economy, however. Much of the cost would have been necessary anyway.

9.3.2.2. Otto Cycle Engine Efficiency

All heat engines perform four basic functions: pumping (compression), heat addition, heat expansion, and heat rejection. Recall from thermodynamics that the efficiency of a heat engine is determined by the difference between its highest temperature of operation (combustion) and its lowest temperature (exhaust). In practice, the highest temperature cannot be greater than engine materials can withstand; neither can the exhaust temperature be lower than the temperature of ambient air since heat transfer occurs only from a hotter to a cooler body. Efficient engine operation is complicated by the limitations of available fuels and the formation of pollutants. Several alternatives to the spark-ignition internal combustion engine have been proposed, including the diesel, the gas turbine, the steam turbine, and the Stirling engine. Each has advantages and disadvantages.

If all the energy in a gallon of gasoline were converted to work, it would provide the equivalent of 200 miles per gallon in, say, a 1971 Ford Pinto. But the theoretical maximum for an Otto cycle engine is only about 57 percent, and so under ideal conditions, would provide "only" 116 miles per gallon in a Pinto. In practice, a 1971 Pinto delivers only about 16 mpg.[4]

American Otto cycle engines are oversized in order to deliver peak power for acceleration, although peak power is used only about one percent of the time of operation. Operating at less than peak power causes energy losses in three major ways. First, mechanical friction in a larger engine is higher. Second, the fuel–air throttle is partially closed at less than peak power, so more work is required to pump the mixture past the throttle into the intake manifold. Third, more of the heat energy in combustion is lost to the cool cylinder walls during partial load operation. Since only about 20 horsepower is required to run a car at 60 mph, it is unnecessary to install 100 horsepower engines to achieve a high velocity. Down-sizing engines from 100 horsepower levels will result in loss of power for acceleration. These problems can be ameliorated by use of stratified charge fuel injection, turbocharging, better transmissions, and electronic control.[4]

The stratified charge injection engine is virtually identical to today's Otto cycle engines, except that the fuel and air are not fed into the cylinder together. Air can be pumped in as with Otto engines, though minus the throttle and therefore with a good deal less work, and the fuel can be injected separately after the air is compressed. Fuel can be injected so as to provide a rich strata of fuel at the top of the cylinder which is ignited first, and with a leaner mixture below this strata which burns uniformly as the piston is pushed downward. This mixture, varied according to the need for acceleration, allows the engine to operate at partial load more efficiently, perhaps 30 percent more efficiently than the same engine without stratified charge. Because an expanding volume of air can hold heat at a lower temperature, fewer nitrogen oxide compounds are produced. A few stratified charge engines are in production today.

Turbocharging is a way of putting the energy of exhaust gases to work. A turbine is placed in the exhaust gas stream for use when acceleration in the car is desired. The turbine compresses a charge of air and rams it into the engine to provide oxygen for more rapid combustion. Though this device adds cost and complexity to an engine, the size of the engine may be reduced since a higher peak power output may be obtained.

Electronic control would add much efficiency to the Otto cycle engine. With fuel and air sensors controlling mechanical devices such as fuel injection, precise quantities of fuel and air could be placed at the optimum place and in the optimum time in the engine. Partial load operation could be much more efficient, especially in city traffic. Unneeded cylinders or even the entire engine could be shut down at stop lights and restarted instantly. Acceleration could be much more effective with smaller engines. Fortunately, electronic control can be affordable.

In addition to transmissions, to which we return after this section on engines, other improvements can be made. Engine friction can be reduced by the use of higher grade motor oils. Claims of five percent improvement today with the use of these blends are valid, though different engines and different oils may obtain different results. Future developments of special coatings for cylinders, lighter engine blocks from magnesium, and other improvements may make the internal combustion spark-ignition engine competitive for many years. With the use of engineering measures such as the appropriate matching of engine-to-power requirements and enhanced combustion control, the Otto cycle engine can be relatively clean and efficient in the near term. This means that we need not wait for development of some uncertain surrogate. Like the technology for producing synthetic fuels, the technology of alternative engines may not work out as well as some would hope. The technologies for improving the Otto engine are available today and can and will be applied to reducing oil imports. With the use of stratified charge technology, Otto cycle engines can even burn the lower grade fuels of the future—fuels with very low or no octane rating.

9.3.2.3. Diesel Engine Efficiency

The chief advantage and disadvantage of the diesel engine is its high compression. It operates similarly to the stratified charge engine in that air is compressed during the third stroke of the engine, and diesel (that is, distillate) fuel is added at the time the piston reaches the top of the cylinder. In this way, the octane limitations on compression are avoided. Diesel fuel will ignite without a spark and burn rapidly as it is injected, provided it has a high Cetane number.* In most other respects, the diesel engine is similar to the Otto cycle engine. The advantage and disadvantage stem from the same aspect. The higher temperature operation made possible by a compression ratio of 14 to 22 (usually above 20, compared to an average of 8 to 10 for gasoline engines) provides better fuel economy, but produces much more nitrogen oxide. In addition, it is difficult to adapt the catalytic converter to reduce NO_x emissions from diesel engines because of the chemistry of diesel combustion. The fuel economy advantage of diesels over gasoline engines, even at partial load, ranges from 20 to 35 percent. But NO_x emissions for diesels remain above one gram per mile, more than twice the original Clean Air Act standard of 0.4 gram per mile which is to be required in the 1980s. Turbocharging can reduce the compression ratio in diesels without loss of fuel economy. But the compression ratio in diesels at present is above the optimum of 15 because of the difficulty of starting diesels in cold weather. Other problems with the diesel include greater noise, smoke, cost, weight, and the possibility that certain particulates it produces are carcinogenic.

* The Cetane number is a measure of the ignition properties of diesel fuel. It is the percentage by volume of Cetane, or $C_{16}H_{34}$, in a mixture with alpha-napthalene, or $C_{10}H_7CH_3$.

9.3.2.4. Rankine Cycle Engine Efficiency

Interest in alternative engines such as the Rankine, or steam cycle, came about initially in response to the need for cleaner engines. Now, there is interest in engines which would not require gasoline but could use whatever fuel was available. A Stanley Steamer type automobile, it goes without saying, is feasible. Fuel of any type can be used to heat water in pipes to make steam which provides motive force by turning a turbine. In practice, though, the Rankine cycle is limited in both efficiency and applicability in automobiles. Efficiency is limited because heat must be transferred through pipes, and with current materials, the highest temperature which may be used is lower than that which may be achieved in an internal combustion engine.* And unless the steam is vented to the air, an arrangement which would require the transport of large quantities of water, it must be condensed. Condensers, or heat exchangers, would be large, expensive, and carry great weight penalties. As a consequence, the Rankine cycle engine is not considered a plausible alternate engine.[9] This assessment might be altered, however, with advances in the use of alternate working fluids.

9.3.2.5. Brayton Cycle Engine Efficiency

A Brayton cycle engine is a gas turbine engine. Air and fuel are ignited in a turbine which turns a shaft that provides motive power and compression (via a second turbine) for the intake air. When operating at full load, gas turbine electric generators are very efficient. At partial load, however, gas turbines are less efficient than gasoline engines. Emissions are low and fuel flexibility is great. Efficiency would be even greater with the perfection of the manufacture of ceramic materials which can withstand high temperatures (but are brittle) for use as turbine blades. The cost of producing gas turbines would be lower than Stirling or Rankine engines, probably, but the issue is whether their greater efficiency, which could only be achieved in cars with the use of advanced transmissions to enable operation at full load, would be worth the expense of switching production from the very cheap Otto engines. Otherwise, the chief advantage of Brayton cycle engines in cars would be fuel flexibility, and even this advantage might be eliminated in comparison with advanced stratified charge engines.[9,10]

---

* High temperature gases in a gasoline engine are "quenched" before they touch the cylinder wall or piston head.

9.3.2.6. Stirling Cycle Engine Efficiency

The Stirling cycle engine is a sophisticated engine invented in 1816 by the Scottish minister Robert Stirling. It is an external combustion engine which can be 40 to 60 percent more efficient than the Otto cycle engine.

The engine is complex, and can be designed in a number of ways. Basic to its operation are a burner which transfers heat through a head to a gas, such as hydrogen, cylinders and pistons which contain and operate with the expansion and contraction of the gas, and a heat recovery device called a regenerator.[10] Its advantages include efficiency and fuel flexibility. Because it is an external combustion engine, it has the additonal advantage of low emissions. The chief disadvantages are high cost of manufacturing because of its complexity, and the severe technical requirement for a material that can withstand the high temperatures necessary for high efficiency. The material through which heat is transferred must be very heat resistant. In effect, the Stirling engine is limited by materials in a way similar to the Rankine engines.[10]

9.3.2.7. Alternative Engines

Table 9.2 summarizes the attributes of the alternative engines we have discussed. Of the alternatives to internal combustion engines, the Stirling cycle appears to have one inherent advantage over the Rankine and Brayton cycle engines: the Stirling would be far more efficient than either of these two at partial load. This advantage would be reduced with the application of advanced transmissions. The estimated cost and complexity of the

Table 9.2. Alternative Automobile Engines

| | Efficiency in percent | | | | Comments | |
| | Full load | | Partial (10%) load | | | |
Engine type	1980	1990	1980	1990	Emissions	Cost
Otto	25	37	15	23	Acceptable with controls	Very low
Diesel	25	40	18	32	No_x unacceptable; carcinogens?	Moderate
Rankine	20	30	15	22	Very good	High
Brayton	25	44	8	25	Very good	Moderate to low
Stirling	30	42	28	38	Very good	High

Source: Adapted from Reference 10.

Stirling engine is a major disadvantage. None of these alternatives is a practical alternative within the next 20 years, because the steps of research and development, the necessary retooling of the manufacturing industry, and then the penetration of the market each might take 10 years.

The spark-ignition, internal combustion engine can be improved by 20 to 50 percent above 27.5 mpg during the 1980s. A 37 mpg gasoline engine car now is cost-effective; a 50 mpg car would be also if the external costs of the national security risks of imported oil and the environmental risks of alternate fuels are considered. Thus, while a new engine might make it possible to use larger cars with greater economy and lesser pollution, an alternative engine for smaller cars may not be necessary. But that is not to say that the present automobile is adequate or that improvements will come naturally. It is our belief that improvements that are needed in terms of national security as well as economics will not come without stringent, technology-forcing government standards.

9.3.3. Transmissions and Fuel Economy

The potential savings in fuel use in engines described in Section 9.3.2. above were based in part on better matching between engine speed and power requirements. The closer the engine is to being fully loaded, the more efficient it is. Transmissions in current use do this job poorly; however, a manual four-speed transmission will usually provide a 10 percent energy savings over automatic transmissions. Ninety percent of all American cars made in 1978 were equipped with automatic transmissions.[3]

The Continuously Variable Transmission (CVT) offers fuel savings of up to 50 percent[4,11] (see Figure 9.3, which describes fuel economy increases for a CVT as a function of speed). Such a transmission would have a large number of gears. Transport trucks have a larger number of gears than cars in today's machines because of the fuel economy the larger number represents. The more closely the transmission can match the minimum speed of the engine to the desired speed of the vehicle, the less pumping, friction, and quenching losses there will be. The CVT would change gears automatically in as small an increment as possible and as close to con- tinuously as possible in order to keep engine speed constant and low, no matter what the velocity of the car. The shifting would entail many, many gears which would be changed hydraulically. The CVT is simply another technology which was not deemed cost-effective as long as energy was trivially cheap.

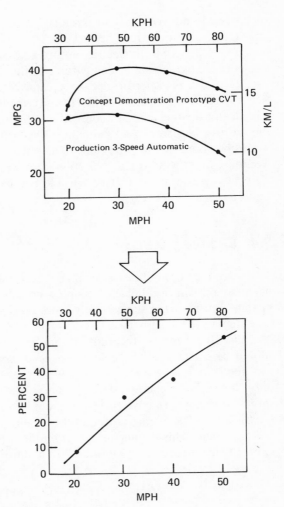

Figure 9.3. Fuel economy improvement due to a Continuously Variable Transmission. Source: Reference 11.

9.3.4. Alternative Automobile Systems: Commuter Cars

Electric or hybrid systems offer a radically different solution to automotive problems. Batteries, of course, limit electric vehicles in range and in speed. But one major fact of transportation life illustrates the potential for short-range vehicles: the fact that 70 percent of all automobile use consists of trips averaging less than 10 miles.[6] Most of this driving is in urban areas where reduced emissions would be worthy of some credit. Most such trips do not require luxury vehicles, and in fact involve an average of only 1.2 persons per trip. The fuel efficiency of electric cars, or hybrid cars which would have both electric motors and conventional heat engines for longer range use, would be comparable to the Otto cycle engine currently in use. It would operate on electricity, and electric conversion is not particularly efficient, especially when the intermediate step of battery storage is added. But the fuel could be something other than oil and would thus offer the positive benefit of fuel switching. The cost of energy to the consumer would be approximately the same. The impact on electric utilities capacity would be slight with as many as 20 million electric cars as long as batteries were charged at night (off-peak). Electric cars could thus be relatively desirable "commuter cars."

9.4. Changing Behavior

Since the automobile can be made safer and more efficient, its use will probably remain the choice of people who can afford it, unless alternative systems can be made substantially more convenient. People generally will take steps to reduce driving time when it exceeds 90 minutes per day.[5]

The first question regarding incentives is, "What mode of transport should be encouraged?" Table 9.3 gives an ambiguous answer. Note that van pools are the least energy intensive form of passenger travel. They are only 40 percent as energy intensive as new heavy rail transit. Full buses are shown to be very effective energywise; but in practice, half-empty buses are not efficient. The average automobile passenger requires more than 10,000 BTU per mile. This is four times that required for van pool passengers and almost twice that for carpool riders.[1] In a major emergency, we will survive

Table 9.3. The Efficiency of Passenger
Transportation Modes

Mode	BTU per passenger mile
Single occupant automobile	14,200
Average automobile	10,200
New heavy rail transit	6,200
Carpool	5,500
Commuter rail	5,000
Bus	3,100
Van pool	2,400

Source: Reference 1.

primarily by forming carpools. Carpooling alone could make up for the loss of 10 percent of foreign oil imports.

Clark Bullard[12] has suggested a policy for encouraging the use of fuel-efficient transportation options. The policy centers on the use of the federal excise tax on fuel for road repairs and other transportation uses and the need to raise this tax from its level of $0.04 per gallon. This tax has not been increased since the 1950s and has been greatly reduced by the effect of inflation. Bullard proposes that we increase the tax and provide the revenues for road repairs, mass transit, etc., to states, provided the states match each dollar granted with $0.20. This five to one ratio contrasts with the current ratio of nine to one. "Bullard's Concept" would allow each state the opportunity to "earn its way back" to the nine to one arrangement by achieving certain levels of increased transportation fuel use efficiency. The state can build subways, promote carpools subsidize van pools, operate buses, or do nothing. But the governor of each state would have a powerful incentive to achieve fuel effectiveness. At the same time, the consumers of transportation could be reaping the benefits of cheaper, faster, cleaner travel.

Automobile consumers might change their purchasing habits, too, if effective, inexpensive alternatives were available. The electric commuter car, or, better, a small, 75 mpg, gasoline powered vehicle might be fine, except when a six passenger station wagon capable of pulling a trailer is needed for vacation, for example. Here we have the alternative of buying different cars for different needs, one (or two) for work, and one for long-range travel. Or, automobile dealers might sell a-few-thousand-miles-per-year use of a station wagon to a new commuter car purchaser. The dealer would own the long distance car and make it available to customers on a reserved basis. Such a practice would reduce the burden of owning several cars. Renting, of course, is another, similar option, as is condominium ownership of less-frequently used machines.[13]

Changed behavior, therefore, need not mean reduced convenience, pleasure, or utility. It may simply mean inventing new institutions to provide our needs under new circumstances.

Enforcing speed limits is an important fuel economy option. Aerodynamic drag increases as the *cube* of vehicle speed. Driving a car 50 miles per hour instead of 55 mph can save 10 percent on gasoline use. But many states do not enforce the speed limits and thus cost thousands of lives in addition to thousands of barrels of oil per day.[6] The Energy Policy and Conservation Act of 1970 requires that states must enforce the 55 mph limit or face the loss of federal highway funds. This highway funds cutoff option should be exercised against noncomplying states, especially in regard to transport trucks which seemingly evade the law at will and at considerable risk to highway safety.

9.5. The Truck

Transport trucks use 2.5 quads of oil annually. They are involved in one out of every 10 fatal highway accidents. And, despite all the recent romance and advertising that America's needs move by truck, only 22 percent of all freight is carried by truck. Barges carry 41 percent, and rail moved 37 percent. Table 9.4 summarizes these statistics in relation to all other modes of moving freight and shows the relative efficiency of each.[1]

9.5.1. Improving Truck Transport Efficiency

One basic shortcoming of today's transport trucks is their lack of aerodynamic design. Trucks are shaped like boxes. A rounded surface on both the front and back would reduce air drag. Cowling around wheels would add efficiency and reduce highway noise.

Truck engines can be improved also, despite the fact that the fleet consists largely of efficient diesel engines. A "bottoming cycle" can be added to trucks which would increase fuel efficiency by 15 percent. A bottoming cycle would involve capturing the waste heat in exhaust for making steam to turn a turbine in a Rankine cycle engine. The extra weight would be tolerable. The additional power would simply be added to the crankshaft. Payback time has been estimated to be on the order of one year. Other devices like declutching fans, turbochargers, etc., could increase fuel efficiency in trucks by 50 percent.[11]

Table 9.4. Freight Transportation in the U.S.:
Modes and Their Fuel Efficiency

Mode	Percent of freight carried[a]	BTU per ton mile of freight
Pipelines	(excluded)	450
Railroads	37	700
Waterways	41	700
Trucks	22	2,800
Air	1	42,000
Total	100%	

Source: Reference 1.

[a] Figures do not add to 100% due to rounding errors.

9.5.2. Encouraging Shifts to Rail

The major potential for freight transport energy savings lies in decreasing (or at least discouraging further increase of) the share of freight carried by truck. Currently this share is increasing to the detriment of the railroads. Policies which suggest themselves include:

o   A user tax on truck freight moved greater than a certain distance (perhaps 100–300 miles) to discourage truck use where rail service is adequate.

o   Raising existing vehicular weight taxes both to discourage truck transport in relation to rail and to help recover the enormous cost of damage to roadways from trucks.

o   Incentives such as a waiver of some of the above disincentives for truckers who use the rail piggyback service (and/or allowing a larger participation of rail-owner trucks in this market).

o   Enforcement of the speed laws (including abolition of radar-foiling devices).

Railroads, of course, are in trouble. Other policy decisions may affect truck use relative to railroads. Increased truck weights will increase the competitiveness of truck transport at the cost of rail. This result should be avoided, as should the damage to highways from heavier trucks and increased danger to truck drivers and other motorists. Damage to roads increases exponentially with truck weight. A truck weighing 73,000 pounds, for example, will do as much highway damage as 6,000 cars, while a truck weighing 80,000 pounds, will do as much damage as 9,600 cars. Safety is diminished, of course, in proportion to the decline in control extra weight causes.

A closely related concern is that of truck length. There is some effort underway to extend the legal limit on truck trailers from 40 feet to 45 feet. In addition to road damage and safety concerns, only one trailer longer than 40 feet can fit on a rail car for "piggyback" service on trains. Two are normally carried when 40 feet in length.

It has been standard operating procedure for railroad executives to bleed railroads of their assets for investment in more profitable ventures. Union featherbedding and interstate regulatory practices have added to this trend by cutting the profitability of railroads. Consequently, the railbeds and rolling stock have deteriorated for many years. Along many stretches of rail in the eastern U.S., the maximum safe speed is 15 miles per hour. Tennessee rails averaged one derailment per day in 1978. Boxcars sometimes spend days in switchyards due to inefficient switching systems.

Controlled lease rates for some rolling stock have been held so low that users frequently utilize them for storage. Railroads, until recently could not be flexible enough with rates to compete with truckers. European rail stock, with more appropriate rates, travels about twice as far per year as does American stock. Unit trains have been applied to coal and other bulky commodity hauling to circumvent this last problem and reduce costs. The purpose of the unit train is frustrated, unfortunately, because off-loading facilities are inadequate. Unit trains are frequently broken up and sections of empty cars returned separately for this reason.[14] Trains are energy efficient and already haul a sizeable portion of U.S. freight; however, if they are to keep their existing market share, much less expand to capture some of that of trucks, something on the order of a national rails rescue policy may be needed.

The subsidies that traditionally have gone to waterways might be a good place to look for support for the nation's railroads. Barge traffic has been competitive because the U.S.'s taxpayers have *built and operated* lock, stock, and canal the entire inland waterway system at an enormous financial (and environmental) cost. The subsidies, in fact, equal the total expenditures of the water transportation industry or about $1 billion per year. The Carter Administration's National Water Policy sought to have waterway users bear some of the burden of construction and operation, but the plan was frustrated by porkbarrel politics. Given the efficiency of railroads which slightly edges barge transport, conceding the vast environmental destruction of dredging whole rivers and filling hundreds of valleys with spoil to create navigable inland waterways, and recalling the need of the railroads, it would seem good public policy to abandon certain water projects and spend the money upgrading rail transportation.

9.6. Airplanes and Energy

Technical and behavioral improvements are needed in the use of airplanes. As long as the value of labor is high, the use of time-saving airplanes is going to remain high. Policies should therefore be directed more at improving passenger and freight miles per gallon than at discouraging their use. Insofar as technical improvements are concerned, these will be diminished to the extent that the profitability of the carriers is decreased. Standards for fuel efficiency should be set for new engines as well as plane design, and incentives (both carrot and stick) should be applied to increase the load factors of airplanes. Policy could include penalties for no-shows, that is, a charge to passengers who make reservations but fail to show up.

9.7. Summing up the Savings

Transportation energy conservation should be our highest priority. Yet, except for the easy goal of 27.5 mpg for new passenger cars by 1985, this sector, which uses 20 quads of oil per year, has been the target of very little action.

Automobile energy consumption could be cut in half in the next 15 years without major new technology. To do so would mean a reduction of five quads annually, or 2.5 million barrels of oil per day, and the reduction could be achieved by requiring automobiles to average 37 mpg. Indeed, the cost of achieving a 40 mpg or more average fleet efficiency should be no more than $80 billion.[7,13] This expenditure (discounted over the 10-year life of new machining tools, equipment, etc.) would save an enormous amount of energy at a cost of only $4.00 to $5.00 per million BTU. (This translates into a cost of gasoline saved of about $0.50 per gallon.) At the same time, a fleet of automobiles would have been produced capable of burning a less expensive fuel, one that could be made from a variety of sources. If the U.S. government desires to fund any crash energy program, it should be one to increase the fuel efficiency of automobiles above 37 mpg as soon as possible.

Transport trucks are the least efficient mode of freight transport. A savings of 20–25 percent is obtainable with current technology. Since the turnover in trucks is relatively fast, and the energy demands of trucks are vast, trucking fuel efficiency improvements should be pursued aggressively. A modal shift of freight from trucks to rail, to the extent possible, would be wise. We have described policies for spurring such a shift.

Airline energy consumption is a small portion of our total energy (see Figure 9.1) but it is the fastest growing. Policies for increasing (or maintaining) high load factors in aircraft are important.

The CONAES[5] evaluation of future transportation energy demand estimated at 2010 total of 10 to 20 quads, compared to 20 today. The lower level would require government intervention, but could produce desperately needed savings at an acceptable cost.

Chapter Ten

Industrial Sector Conservation

10.1. Unplanned Obsolescence

Reid and Chiogioji described American industrial equipment as probably the most energy-inefficient in the world. Comparisons of energy consumption by major industries among various countries tend to verify this assertion. Energy, of course, is not the only factor of production which managers must consider. Labor, time, materials, and capital most also be conserved. When an industrial plant is built, its builders seek to minimize the total costs of all these factors combined. When most of America's industry was built, however, oil cost $2.00 per barrel, and natural gas cost far less. The Great Embargo of 1973 and subsequent events made many of our industrial plants obsolete.

Our economy requires chemicals, iron and steel, paper, cement, aluminum, food, asphalt, etc., and the production of these goods is energy intensive. The production of these materials is growing rapidly, as well. The energy used for production of chemicals, for example, may increase by 200 to 300 percent by 2010.[1] To conserve energy in new plants is easy; energy-efficient design may be incorporated or even whole new processes installed. But retrofitting, upgrading, and replacing existing industrial capital equipment will be an expensive proposition. Despite the fact that large capital investments will be required, however, the savings in energy costs can be quite attractive. Unfortunately, capital expenditures for energy conservation are approached cautiously by industrial management. Alcoa Aluminum, for example, requires a two-year payback period on such investments.[2] Government policymakers can help assure that we do not discount the future so heavily.

Ross and Williams complained that a major failing of present research programs is that the nation does not spend enough effort rediscovering and reorganizing existing technical knowledge. This complaint may have been prompted in part by the failure of industry to take advantage of major energy saving options such as cogeneration. Industrial cogeneration, the combined generation of electricity with steam for industrial process heating, or for other uses such as district heating, is an option of enormous potential, but one which has been and continues to be ignored for institutional reasons. Of course, there are many lost opportunities for saving energy in industry. Basic housekeeping, simply the repairing of leaky steam pipes, broken windows, and so forth has not yet been fully exploited. Moderate levels of capital investment in heat recovery devices are more than cost-effective. Even very large capital investments, including the scrapping

Table 10.1A. Annual Energy Demand of Energy-Consuming Industries (1975)

Industry category	Annual energy demand (quadrillion BTU)
Chemicals	
Fuel and power	3.0
Feedstocks	1.5
Subtotal	4.5
Iron and steel	3.0
Paper	2.2
Asphalt and cement	1.4
Agriculture	1.2
Food	1.1
Aluminum	0.6
Other	6.0
Total	20.0

Source: Reference 1.

Table 10.1B. Annual Energy Demand of Energy
Producing Industries (i.e., Conversion Losses, 1975)

Industry category	Annual energy demand (losses) (quadrillion BTU)
Electric utilities	13
Oil refining	3
Coal mining	0.2
Oil and gas extraction	0.2
Total	17

Source: Reference 1.

of some existing plants and replacing them with new equipment are frequently feasible, provided management can raise capital. Like homeowners who must somehow find the money to buy attic insulation in order to save a lot more money, industrialists must be able to finance energy conservation. Government can play a key role in assuring that financing is available for industrial energy conservation.

This chapter cannot possibly detail all the opportunities for energy conservation in industry. It can, however, serve to illustrate opportunities and problems which cut across industry categories and to provide examples of what might be accomplished. The greatest opportunities, of course, lie in those industries which consume the largest amounts of energy today.

10.2. Industrial Demand Characteristics

10.2.1. Current Industrial Energy Demand

A little more than 25 percent of U.S. energy demand is industrial. By industrial, we mean energy consuming industries. Energy consuming industries, the largest of which are, in order of annual energy requirements, chemicals, iron and steel, paper, asphalt and cement, food, agriculture, and aluminum, use 20 quads each year. Energy producing industries, which are, in order of annual energy requirements, electric utilities, oil refineries, coal mines, and oil and gas wells, use almost 17 quads each year. Thus it is evident that we use 22 percent of our annual energy budget just to produce energy. By far, the largest part of this cost to our energy budget is electrical generation and distribution losses. The use of electricity costs 13 to 14 quads per year. Oil refining consumes three quads yearly, and coal mining and petroleum production require small amounts also. Tables 10.1A and B summarize these details.

10.2.2. Industrial Demand and Thermodynamics

It is instructive to observe the energy forms, or quality, of industrial energy demand. Electricity, even as expensive as it is relative to other fuels, provides industry with about 20 percent of its purchased energy. Steam is the form in which a third of all industrial energy is applied. Nearly half of all industrial energy is consumed as hot gas. Table 10.2 provides a breakdown of industrial energy consumption by industry category and energy form (hot water, steam, and hot gas).[3] These figures are instructive to efforts to match energy supply and demand in the most thermodynamically optimum way.

Table 10.2 points out both opportunities and constraints. One opportunity not immediately evident in the table is that a large quantity of the steam and hot gas or direct heat applied in industrial applications is used for relatively low temperature heating in the 100°C to 150°C range. Much of this energy could be supplied by solar devices,[4] albeit at considerable cost of capital.

One opportunity that should become immediately apparent is the amount of steam which might be available for industrial cogeneration. Because the potential for cogeneration is so great, we devote a large portion of this chapter to industrial cogeneration, the installation of electric generating turbines on both existing and new industrial boilers. These boilers should be considered literally as energy mines from which the U.S. could obtain an additional two to five quads annually.[1,5]

An energy-saving possibility which may not be obvious but is implicit in Table 10.2 is that a lot of high grade energy is applied in processes that could use lower grade heat. Certain processes such as the annealing of metals or the reduction of metal ores certainly require high grade energy (high in a thermodynamic sense). Many applications, however, such as drying or evaporation need not consume such intrinsically valuable fuels as natural gas. The high temperature combustion of natural gas to perform what could be low temperature tasks is a serious thermodynamic mismatch. Thermodynamics can, in such cases, instruct us in how to conserve energy.

Thermodynamic matching of energy supply and demand suggests the possibility of energy cascading. Cascading involves arranging flows of energy in industrial plants in order that processes which may require high grade heat receive first a flow of steam, hot air, or hot water. The waste heat rejected from a first application can then be applied to a process requiring lesser grade heat, and so on. The cascading of heat used for burning raw materials to make cement exemplifies the process. From a cement kiln, waste heat can be applied to preheat raw materials to drive

Table 10.2. The Forms of Energy Used in Major Industries[a]

| Industry | SIC[b] | Percentage of process heat requirements | | | | | Electrical requirements as a percent of total energy requirements |
		Hot water 100°C	Steam 100–200°C	Steam 200–350°C	Direct heat or heated gas	Total	
Rubber	30	0	0	36.3	63.7	100	29
Machinery (except electrical)	35	16.3	17.6	26.7	55.2	100	29
Electric machinery	36	0	0	0	100	100	39
Paper	26	14.2	60	6.3	20.5	100	17
Apparel	23	NA	NA	NA	NA	100	NA
Chemicals	28	1.2	43.4	16.9	38.4	100	18
Fabricated metals	34	0	0	0	100	100	26
Food	20	14.5	4.7	0	80.9	100	16
Textiles	22	55.1	15	0	29.8	100	33
Primary metals	33	4	1.9	0	94	100	29
Stone, clay, glass	32	1.5	2.5	2	94.1	100	8

Source: Reference 3.
[a] NA = not available.
[b] SIC = Standard Industrial Classification.

off water otherwise requiring energy in the kiln itself. Alternatively, waste heat could be supplied from another factory altogether.

Heat recovery with or without energy cascading is an essential element of industrial energy conservation. It involves the use of heat exchangers in hot gas exhaust stacks to "recycle" heat and the use of other similar devices.

10.2.3. Industrial Energy-Consuming "Appliances"

The complexity involved in the evaluation of industrial processes at first seems staggering. The category chemicals, for example, may be aggregated on a level that includes industrial inorganic chemicals, nitrogenous fertilizers, organic fibers, alkalies, plastics, etc. At this level of categorization in industry as a whole, there are approximately 450 industries, each with 5 to 15 process steps. Thus, it would seem that the energy conservation analyst would have to study 5000 different processes in order to comprehend the potential for energy reduction in industry.

Fortunately, there is a manageable number of common unit processes which are applied repeatedly. These processes include high temperature energy applications, such as annealing and baking, and relatively low temperature heat applications, such as distillation, evaporation, washing, and drying. No more than sixteen such energy uses account for more than half the energy used in industry in America.[6] In a manner of speaking, then, there are 16 "appliances" which consume half of the energy used in manufacturing, or between seven and eight percent of all the energy used in the U.S. By examining the energy conservation potential in these "appliances" we can make judgements about the overall potential for saving energy in the industrial sector.

Table 10.3 lists 10 of the energy-consuming industrial appliances we refer to and describes them in terms of the quantity of energy each consumes, the percent of total industrial energy consumed, and the thermodynamic efficiency of each. Note especially the extremely poor (second law) efficiency of the low temperature processes. Dryers, for example, are less than .3 percent efficient.

In Section 10.3 we return to these appliances for an examination of their conservation potential. But first we should examine the past and present trends in energy use in U.S. industry.

10.2.4. Historical Trends in Industrial Energy Consumption

Time series is a statistical term which refers to data collected at intervals over time, such as annual won–lost records of basketball teams.

Table 10.3. The Efficiency of Major Industrial Energy Applications in the U.S.
(Based on 1974 Data)

Industrial "appliance" (process)	Annual energy demand (quadrillion BTU)	Percent of industrial demand	Second law efficiency (percent)
A. High temperature processes			
Furnaces	1	5	49
Reactors with preheaters	3	15	43
Cokers	1	5	80
Annealing units	0.4	2	NA[a]
Ovens and heaters	0.3	2	23
Subtotal	6	30	
B. Low temperature processes			
Distillers	2	8	4
Evaporators	0.2	1	6
Sterlizers	0.2	1	5
Dryers	2	8	0.3
Washers	0.2	1	3
Subtotal	5	20	
Total	11	50	

Source: Reference 6.
[a] NA = not available.

A time series analysis of the rate of change in industrial energy consumption shows a surprising and encouraging result. Demand for energy per unit of industrial production has actually been declining for the last thirty years.

Figure 10.1 illustrates this fact. The chart compares an index of industrial production with industrial energy demand in two years, 1947 and 1975. A comparison of the relative changes makes it evident that growth in production has outstripped growth in energy use. The actual rate of decline in the use of energy per unit of output was 1.5 percent per year. Importantly, the price of energy, in real terms, actually fell at a rate of one percent per year over the same period.[1]

Several explanations for the improvement have been offered.[1]

o   The conversion from coal to gas and oil resulted in higher process efficiencies because of the nature of gas and oil and the advantages they offered in particular forms of utilization.

o   Increasingly larger processing units began to make it economical to recover the heat from waste streams.

o   Many new industrial plants were sited in areas with relatively mild climates (more than half over the last 30 years), thereby

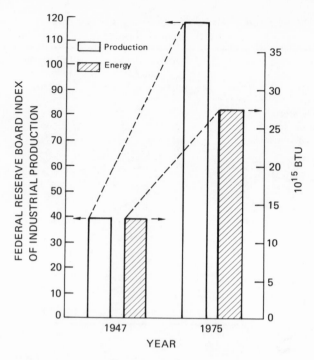

Figure 10.1. Industrial production growth compared to growth in energy use. Source: Reference 1.

reducing space heating loads and heat losses to the atmosphere. The major area of industrial expansion, the Gulf Coast, had the double advantage of a mild climate and low energy costs.

o A number of energy-intensive industrial installations began to meter and manage their energy consumption more carefully.

o New technologies developed and expanded at an increasingly rapid rate.

o New energy-saving devices (e.g., gas turbines, finned-tube heat exchangers, cryogenic turbo-expanders, and high temperature steam superheaters) were rapidly developed and brought into use.

Although much of the potential of the above items has probably been realized, improvements in industrial operating practices and the development of new process technologies are likely to continue, especially in an era of climbing energy prices.

10.2.5. International Comparisons

Comparisons of energy consumption across national boundaries are filled with problems arising from differences in climate, resource availability, prices, etc. But comparing American and European applications of energy in specific industrial processes for fuel efficiency should be instructive. Table 10.4 offers several such comparisons.

The U.S. used more energy per unit of output than any other nation in each of the four industry categories shown in Table 10.4. In steel production, the U.S. uses 50 percent more energy per ton than Italy, which, among the five countries evaluated, was the most efficient steel producer.[7] In cement manufacture, the U.S. uses 80 percent more heat than West Germany, which ranks most efficient in this category, and 45 percent more electricity in cement production than England.[8] Similar comparisons hold for plastics and chlorine production. The U.S. possibly does have the least efficient industrial plants in the world. Such a state will further disadvantage the U.S. in world markets, as it has in the steel industry. On the other hand, such excess energy consumption for the production of goods indicates a promise that energy productivity can be vastly improved.

10.3. Upgrading Industrial Processes

There are three levels of effort in increasing industrial energy productivity: housekeeping, or noninvestment, operational improvements; a moderate amount of capital investment; and very large capital investments. After summarizing recent housekeeping gains, we will consider the more difficult but more rewarding levels of effort.[9]

Table 10.4. International Comparisons of Industrial Energy Use in Specific Applications

Industrial application	Relative energy use in percent (100% = best)[a]				
	Netherlands	West Germany	Italy	England	U.S.
Cement manufacture					
Heat	NA	100	110	155	180
Electricity	NA	110	115	100	145
Plastic (PVC)	101	NA	105	100	125
Steel production	135	110	100	130	150
Chlorine production	125	NA	115	100	130

Source: References 7, 8, 9, and 10.

[a] NA = not available.

10.3.1. Housekeeping

Housekeeping is the basic minimum required for effective energy management. Generally it should include employee awareness programs, energy auditing, shutting down production units not in use, lowering and raising thermostats as demand varies, installing low-cost insulation, and establishment of corporate energy conservation goals coupled with top management review of results. The importance of commitment to energy conservation by management cannot be overemphasized.[9]

The Energy Policy and Conservation Act (EPCA) of 1975 set voluntary guidelines for industry to try to meet by 1980. The actions promoted were essentially housekeeping functions. Actions called for include fixing steam leaks, turning off machinery not in use, replacing broken windows, disconnecting the coffee pot over the weekend, and so forth. The goals were not trivial, however, for they ranged from a goal of nine percent for primary metals, a very energy-intensive industry, to 24 percent for fabricated metal products.[9]

Repairing simple steampipe leaks can be quite rewarding. Steam under 200 pounds of pressure leaking from a hole only one-tenth of an inch in diameter can waste more than $1200 worth of energy in a year's time. Repairing several such leaks could easily pay the salary of a full-time energy manager.[10]

By early 1977, industry categories had already made progress toward achieving the 1980 goals:

o   Chemicals industries had saved nine percent out of a goal of saving 14 percent.

o   Petroleum refiners and machinery manufacturers had exceeded their goals of 17 percent and 12 percent, respectively.

o   Food processors and packagers had reduced their energy use by 11 percent, relative to a goal of 12 percent reduction.

On the other hand, the steel industry had achieved only half of its goal, and the fabricated metals industry had achieved only four percent savings despite a goal of 24 percent. A small percentage of a large number, such as steel industry energy demand, is still a large number, however. A four percent overall energy reduction in the iron and steel industry can be achieved with virtually no investment and save .1 quad per year worth about $250 million.

10.3.2. Capital Investments for Energy Conservation

Opportunities abound for small capital investments which will save large amounts of money by reducing industrial energy bills. As similar "appliances" are applied across many industry categories, many similar energy conservation investments can be made. Waste heat recovery devices, for example, can be applied in dryers, kilns, furnaces, compressors, and so on. Insulation investments, small, medium, and large, can be applied throughout all of the industrial sector. Replacing equipment poorly matched between energy or power requirements and capability can be profitable. The list is long, and perhaps a specific example will illustrate the potential. Upgrading evaporators is one option that has received detailed analysis.

Evaporation is the removal of a solvent from a solution by vaporization of the solvent. It is similar to distillation, except that (1) it is the unevaporated portion left behind that is the valuable product, and (2) no effort is made to separate out the volatile components of the evaporated fraction, as is done in the distillation of petroleum products (diesel fuel, gasoline, etc.). Fresh water is distilled from sea water; salt is produced by evaporating water. Orange juice is concentrated in evaporators, as are chemicals and pulp products.

Evaporators are manufactured in discrete stages, called effects, through which the feed liquid is pumped. Steam is piped through an effect causing the portion of the liquid with the lowest boiling point to boil off first. Water, for example, boils long before salt will even melt.[11]

The cost of operating a typical evaporator has risen almost 300 percent over the last decade, and all but 27 percent of the increase is attributed to increased steam costs. An evaporator using 33,000 pounds of steam per hour required $175,000 (1980 dollars) worth of steam in 1967 (at $0.50 per 1000 pounds), compared with one million dollars worth of steam (at four dollars per thousand pounds) in 1980. Evaporators use almost .2 quad per year.

Low-cost options for upgrading evaporators include heat exchangers for heat recovery, improved maintenance, and thermal insulation. These investments will save 10, five, and five percent of total evaporator energy use, respectively. Thermal recompression (injecting steam in the evaporated fraction to facilitate the recovery of heat in the vapor) and mechanical recompression are conservation opportunities that require medium to high capital investments, but energy savings of from 45 to 90 percent are possible. Adding extra effects is also a high-cost, high-return option. Adding a fourth effect to a three-effect evaporator will save approximately 25 percent.

Adding a heat exchanger costing $40,000 (1980 dollars) to an evaporator built in 1967 (costing $460,000) and designed to use 33,000 pounds of steam per hour will save almost 21 billion BTU per year per evaporator at a cost of about $0.32 per million BTU.* Adding both a heat exchanger and an additional effect with a total cost of $395,000 will reduce the hourly demand for steam to 21,200† pounds. The cost of energy saved would be $0.72 per million BTU. The discounted rate of return on the two investments are 200 and 80 percent, respectively. Systems using 40,000 pounds of steam per hour (40 million BTU/hr), and using steam at a cost of $3.40 per thousand pounds, can justify a $2 million capital investment and expect a 30 percent rate of return.[12]

10.3.3. Policies for Stimulating Capital Investment in Conservation

Four federal policies for stimulating industrial energy conservation suggest themselves: (1) provision of incentives such as rapid depreciation or investment tax credits for retrofitting existing industrial appliances such as evaporators, dryers, kilns, furnaces, etc.; (2) imposition of appliance standards; (3) a vigorous research and development program geared toward energy production; and (4) provision of incentives for fostering general productivity increases.

Subsidies for industrial conservation were provided in the National Energy Act (NEA). The NEA offers an additional ten percent investment tax credit which generally represents a subsidy of four cents for every million BTU saved. This incentive is not adequate.

Some analysts object to targeted industrial energy conservation incentives altogether. Their objection is that such incentives distort the economics of industrial production in deleterious ways, and that a better alternative is to provide general incentives for increased industrial productivity through capital investment and research and development. We agree that increased industrial productivity should be a national goal, and that this goal could be advanced through more rapid tax depreciation of industrial equipment. But we disagree that incentives targeted for industrial energy conservation are harmful. To the contrary, industry may postpone conservation investments simply because these will not go

---

* Assumes an annual levelized fixed capital charge rate of 15 percent, constant 1980 dollars.

† 90 percent load factor; 880 BTU per pound of steam.

Table 10.5. CONAES Estimated Net Energy Intensity[a]
Reduction in Energy-Consuming Industries (by the Year 2010)

Industry	Energy prices double "B scenario" (percent reduction)	Energy prices quadruple "A scenario" (percent reduction)
Agriculture	15	15
Aluminum	37	45
Cement	37	40
Chemicals	22	26
Construction	35	42
Food	24	34
Glass	24	31
Iron and steel	24	28
Paper	29	36

Source: Reference 1.
[a] Intensity refers to BTU/unit of output.

away. Industrialists may feel compelled instead to make investments in increasing their ability to capture a certain market, for example, the kind of opportunity which might disappear forever. Therefore, we would encourage the provision of targeted incentives for conservation but with a time limit on availability in order to advance their adoption.

Note in Table 10.5 the conservation potential for the nation's largest industrial energy users. Chemical manufacturing could reduce consumption per unit output more than 20 percent; iron and steel, 24 to 28 percent; paper, 29 to 36 percent; and so on. Comparing these estimates, which are from CONAES, with the performance of European countries suggests that the CONAES estimates are, at the very least, achievable.[1]

10.3.4. New Processes

On the short to medium time horizons, it will be retrofitting and replacement of industrial equipment that offers the greatest industrial energy conservation potential. In the long run, brand new ways of doing things could greatly alter energy consumption patterns. Invention has been constantly changing consumption patterns, as a few vignettes will serve to illustrate:

○ The progression from vacuum tubes to transistors and then to large scale integrated circuits has enabled the energy required for various functions in communications and computers to be reduced about a million-fold in three decades.

o   The efficiency of electricity conversion in lighting has increased from less than five percent (incandescent) to more than thirty percent (alkali–halide fluorescent) since 1950.

o   The energy required to produce a pound of aluminum from alumina has fallen from more than 12 kilowatt hours per pound in 1950 to less than five kilowatt hours per pound using the new Alcoa molten chloride process.

o   The energy required to produce low density polyethylene has been halved because of a new process developed by Union Carbide.

o   Research on producing a fire-safe diesel fuel accidentally resulted in a fuel that burns with 10 percent better economy and 50 percent less exhaust smoke.

o   Energy consumed per unit output of product has fallen substantially for all industry over the past decade, as much as 50 percent for many petrochemical projects.

In the next few years, additions will undoubtedly be made to this list. The Alcoa process for making aluminum will use 30 percent fewer kilowatt hours than the Swiss and French aluminum manufacturers whose plants are presently 25 percent more efficient than the average American aluminum plant. Interestingly, the Alcoa process helps eliminate the old problem of fluoride emissions.

Low temperature reduction of limestone may be feasible and replace the current practice of high temperature firing with natural gas in prodigious quantities.

Many industries are amenable to process changes, changes that are not necessarily wholly related to energy conservation, but have conservation as a side benefit. CONAES for instance estimated that aluminum, cement, steel, paper, chemicals, and food industries were amenable to new processes. Researchers caution, correctly we think, that research and development must not focus too narrowly on a problem, but should look at the whole picture, seeking solutions to a broad range of problems.

10.3.5. Resource Recovery

The recycling of scrap, mostly iron and steel and copper, is already practiced in a large way in the U.S. Two-thirds of the nation's iron and steel, and a substantial portion of its copper, are produced from scrap. Yet, missed opportunities for recycling are becoming crises in many metropolitan areas where landfills grow like slime mold. The scarcity of urban land suitable for landfilling, the public resistance to the practice, and the lost resources make resource recovery increasingly appealing.

The average American produces about four pounds of solid waste per day. Municipal solid waste (MSW) is about five percent by weight ferrous materials which have a market value of $25–40 per ton. Sixty percent by weight of MSW is combustible with a heat value of about 5000 BTU per pound.* Recall that paper has a BTU value in this range. MSW can be "refined" to separate out the recyclable metals and fuel, or the whole mess can be dumped into waterwall furnaces and burned. In the latter method, whole engine blocks may pass through the furnace and be collected out of the back side. In the former, magnets, hammermills (shredders), and air or mechanical classifiers separate out recyclable metals, fuels, and refuse.

Tacoma, Washington, for example, will soon shred and classify 600 tons of garbage per day and hopes to sell refuse derived fuel (RDF) at 40 cents per million BTU to pulp and paper mills located nearby. The mills could blend RDF with wood waste fuel. A major obstacle to this technology may be explosive materials such as compressed gas cylinders which would destroy hammermills. Furnaces for burning MSW, fortunately, are designed to withstand the explosive force of a hand grenade.[13]

If it were economical to recycle all the components of municipal solid waste, an amount of energy equivalent to 2.1 percent of the U.S.' annual consumption could be conserved. National legislation requiring the return of beverage containers alone could save .17 quad per year as well as cut highway litter from this source by 80 percent.[14]

Resource recovery requires both governmental encouragement and research and development. The former should be provided by local governments which must find solutions for the waste problem. The latter should come not only from the federal government, but from private enterprise, since resource recovery can be profitable.

Policies designed to minimize the generation of garbage also are needed. In the long run, it may be unfortunate that it is now often cheaper to trash a radio and buy a new one than have the old one repaired, to burn worn tires rather than retread them, or abandon old clunker cars in the junkyard rather than repair them. Making goods more durable and expanding the service sector to extend the lives of products through servicing and repair would have environmental, energy, and probably employment benefits. Perhaps we would require longer

---

* Compare with eastern bituminous coal at 12,000 BTU per pound and western subbituminous coal at 6000–8000 BTU per pound.

warranties on durable goods. Perhaps a deposit fee on television sets and appliances would result in more of these being recycled. Perhaps we will use our cunning so that Tom Lehrer's lyrics will not become prophesy[15]:

> Garbage! Garbage!
> What will they do with the Garbage?
> What will they do when there isn't any room
> For the
> Garbage!?!

10.4. Industrial Cogeneration

Cogeneration is an idea at least as old as the steam engine. It is a relatively simple concept, but one that has become mired in a swamp of institutional constraints. Used in Europe to produce more than a third of all industrial power, cogeneration has declined to four percent of U.S. electric generation. Cogeneration, after automobile fuel economy, may represent our single largest missed energy conservation opportunity and may be a classic example of our failure to meet the energy challenge.

10.4.1. Cogeneration Defined

Cogeneration is the simultaneous generation of electricity and heat or motion for industrial processes or other heating purposes. Several configurations are possible, although three predominate.

Central power stations are cogeneration units when the waste heat of condensing steam is applied to do work. A central station power plant in Sweden, for example, will have the capability not only of generating electricity, but also of providing hot water for space heating to a city 90 miles away from the power plant. Minneapolis/St. Paul, Minnesota, is reconstructing a district heating system using existing steam pipes and a downtown power plant. Steam is more costly and troublesome to transport than hot water, however.

Of much greater potential application is the use of steam generated for industrial processes. Central station plants produce copious quantities of low grade heat. This heat is usually not available where it can be used. Smaller industrial heat applications are more adaptable to cogeneration. Steam may be generated in a boiler for first turning a steam–electric turbine and for secondary application to an industrial process. Alternatively, a diesel or gas turbine engine may generate electricity by turning a turbine attached to a drive shaft and producing steam or hot water for industrial process use by utilization of the waste heat of these engines in

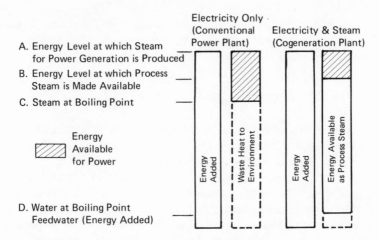

A. Energy Level at which Steam
for Power Generation is Produced

B. Energy Level at which Process
Steam is Made Available

C. Steam at Boiling Point

D. Water at Boiling Point
Feedwater (Energy Added)

Figure 10.2. Energy savings in cogeneration. Source: Reference 16.

a waste heat boiler. Industrial cogeneration plants generate electricity for approximately half as many BTU per kilowatt hour (kwh) as central station power plants.

Figure 10.2 illustrates one reason for cogeneration's high efficiency. Point D on the chart represents the heat in water, or feedwater added to a boiler. Note that to go from D to C, that is, to heat feedwater to steam at the boiling point (100°C or 212°F), requires far more energy than to raise the energy level of steam from C to B or from B to A. B is the point at which industrial process steam may be required—perhaps at 50 pounds of pressure per square inch (psi) and 200°C. Steam at A—say 300°C and 150 psi—may be efficiently used to generate electricity. Extracting work to generate electricity reduces the energy in steam from A to B. But the energy indicated between B to D is still available for industrial applications. Using steam only for making electricity, as we do in the overwhelming majority of power plants, wastes the energy from C to D.

The energy required to go from D to C, in Figure 10.2, is the energy required to go through the phase change from liquid to steam. Expending energy once for process steam and electric generation is far more efficient than undertaking the two processes separately and twice incurring the losses inherent in the water to steam phase change.[16]

The most efficient cogeneration configuration consists of a combined cycle composed of a diesel engine or Brayton cycle engine and a waste heat boiler. A Brayton, or gas turbine engine, may be little more than a jet engine, is used to propel an electric generator. Exhaust gases from the jet are channeled through a steam generator to generate steam to produce more electricity and/or for process applications. Combined cycle systems are not very energy efficient but produce power for approximately two-thirds the cost of central station plants.

10.4.2. The Efficiency of Cogeneration

Generating electricity in a coal-fired, central station electric plant requires about 10,000 BTU per kilowatt hour. Since a kilowatt hour has an energy value of 3413 BTU, the thermal efficiency of electrical conversion is therefore about 33 percent. But to generate steam at a level above that required for most process applications in industry requires only about 4550 BTU per kilowatt hour in a steam cycle plant. The second law efficiency of a steam cogeneration plant may be 40 percent, compared to 32 percent for separate process steam and steam electric generation. For combined cycle (gas/steam) plants, the second law efficiency can exceed 50 percent.

The importance of applying Brayton or diesel cycle systems to cogeneration opportunities should not be underestimated. These systems will produce four to six times as much electricity per unit of thermal energy demand as will conventional Rankine systems. The use of gas or oil in cogeneration could increase the potential of cogeneration from only 20,000 megawatts to 100,000 megawatts. Gas turbine systems associated with the Pressurized Fluidized Bed Combustor, capable of using solid fuels such as coal or biomass, should become available sometime in the 1980s. Natural gas, made more abundant by price deregulation, should provide an efficient transition fuel for use in industrial cogeneration, and its use in this capacity should not be discouraged. Residual oil, low in quality and relatively inexpensive, is also a good fuel for cogeneration. Federal policy, again, should not discourage the combustion of residual oil in industrial cogenerators.

10.4.3. The Economics of Cogeneration

The economics of industrial cogeneration can be impressive. If a minimum steam load exists, say, 50,000 pounds per hour* for gas turbines and 100,000 pounds per hour for steam turbines, cogeneration is economically feasible in most areas. As Table 10.6 indicates, cogeneration can be attractive with low load factors,† even in areas with relatively

---

* 1000 pounds of steam roughly equals 1 million BTU. Therefore, 50,000 pounds of steam per hour equals 50 million BTU per hour, approximately.

† A load factor is the fraction of maximum capacity actually utilized over time in a system.

Table 10.6. Cogeneration Costs vs. Separate Steam and Electricity Generation—
A Sample Comparison

System type	Size	Capital cost ($(1977)/kw)	$/kwh[b]	$/1000 pounds[b] of steam
Nuclear central[a] station	1000 MW	$800–$1000	$0.045	—
Package gas boilers[f] for steam only	80,000 PPH	—	—	$5.00
Steam turbine cogeneration[g]	5 MW	1000[c,d]	0.03	3.25
	30 MW	600[c,e]	0.025	2.50

Abbreviations used: MW = megawatt; kw = kilowatt; kwh = kilowatt hour; PPH = pounds per hour.
[a] From Reference 21.
[b] At 65 percent load factor (.65). One kwh = 3413 BTU. One thousand pounds of steam equals approximately one million BTU.
[c] Costs are allocated between the electrical and steam systems according to the methodology described in Reference 5.
[d] Total steam plus electric capital costs equal $2000/kw (1980 $). Some sources indicate this cost should be much lower, which would be even more favorable to cogeneration.
[e] Total steam plus electric capital costs equals $1200/kw.
[f] Fuel costs are calculated according to standard levelized cost analysis. Backup steam and standby electric charges are added in accordance with Reference 5.
[g] For the Tennessee Valley Authority Region. From Reference 17.

inexpensive electricity. The most valid societal comparison for the cost of cogenerated electricity is the marginal cost of power. As Table 10.6 shows, marginal electricity would cost more than $0.04 per kilowatt hour. Steam generated separately would cost from $3.00 to $5.00 per 1,000 pounds, or $3.00 to $5.00 per million BTU. Cogenerated electricity, though, would cost only $0.03 per kilowatt hour, and cogenerated steam would cost only $2.50–$3.25 per thousand pounds.

10.4.4. The Problems of Cogeneration

Because cogeneration conserves both energy and money, one might anticipate a run on stock in cogeneration equipment. But estimates of the rate of expansion of cogeneration are pessimistic.[18,19] The reasons include those mentioned previously along with others, both technical and institutional. Technically, several requirements impose themselves on the industrial owner:

o The need for additional space for storage of solid fuels such as coal or wood, unless the industry can get an exemption from the Power

Plant and Industrial Fuel Use Act of 1978 to use oil or gas for cogeneration.

o   Additional expertise is required of industry staff.

o   Steam loads in certain industries fluctuate too greatly for economic application of cogeneration.

o   Steam may be used in large quantities for mechanical auxiliaries[5] and thus be limited for electric generation.

It is the institutional problems that have been the most intractable, however. These include:

o   Discouragement by federal regulations such as those imposed by the Fuel Use Act which have the effect of constraining cogenerators to use the least efficient cogeneration systems.

o   Industries' desire to stay away from activities that invite further regulations; similarly to leave the responsibilities of meeting environmental regulations to the utilities.

o   Utilities' desire to have the most reliable energy supply. Industrialists may be unwilling to commit themselves to use of an unfamiliar technology.

o   A failure of the state public utilities commissions to force consideration of cogeneration.

o   *Lack of strong financial incentive.* Energy costs usually equal about 10 percent of industries' total costs. A 10 percent energy cost increase would thus add only 1 percent to total costs, a relatively small inducement to change.[17]

The Public Utilities Regulatory Policies Act (PURPA) of 1978 offers a powerful incentive for industrial cogeneration, however. PURPA requires electric utilities to pay industrial cogenerators the marginal cost of cogenerated electricity. This means that industries can sell excess power generated on-site for a price equal to the cost of power from, for example, a new nuclear plant. Since the cost of cogenerated power will usually be less than that from a new central station unit, then cogenerators will be motivated to generate as much surplus power as possible. PURPA should also motivate utilities to enter into joint ventures with industries to cogenerate. A prototypical model of such an arrangement is described in the next section.

10.4.5. A Model for Industrial Cogeneration

A good example of the cunning and determination required to make conservation work is a recently negotiated cooperative arrangement between the Eugene (Oregon) Water and Electric Board and the Weyerhauser

Corporation. Six months of contractual negotiations were required, but the result was a model for the nation. It is called the Utility Industrial Energy Center.

In the early 1970s, the Pacific northwest experienced power shortages because of drought, and Weyerhauser, in order to assure a power supply for its Eugene pulp operation, ordered a 52 megawatt steam-electric turbine to be connected to three wood-waste (black-liquor) boilers. Oil was to be burned to raise the temperature and pressure of the steam to a degree sufficient for electrical generation. When the turbines arrived, however, Weyerhauser found it did not need them because it had since rained. It then contacted the Eugene Water and Electric Board (EWEB) to discuss a cooperative use of the turbines. Weyerhauser first wanted to sell EWEB undivided interest in the plant, with the utility disposing of the power except when Weyerhauser needed it. Weyerhauser's plant consumes about 50 megawatts which it purchases from EWEB. EWEB purchases 80 percent of its 500 megawatt load from Bonneville Power Administration (BPA), and sells Weyerhauser its 50 megawatts for about five mills* per kilowatt-hour on an interruptible basis. Legal and political constraints, however, would not allow EWEB to enter into an undivided interest holding, or to allow Weyerhauser emergency use of the 50 megawatts generated on-site but normally exported to the utility grid. The power, both realized, would have to be sold off-site becasue on-site cogenerated power would cost 18 mills per kilowatt, considerably more expensive than BPA hydropower.

EWEB was motivated (mainly due to the energy and foresight of Herbert Hunt, a vice-president of the small but historically innovative utility) to secure additional capacity for future needs. Hunt saw in this situation an opportunity to accomplish this goal by buying the turbines, which he installed on land rented from Weyerhauser adjacent to the boilers, and exporting the power to Pasadena, Burbank, and Glendale, California, eight hundred miles south. Thus, the Oregon utility's future electric capacity needs are being financed and will be available when needed locally. In order to finance the project which included the costs of the turbines and auxiliary equipment in addition to the considerable costs involved in retrofitting of the boilers, the utility issued highly rated bonds for $7.2 million dollars.

The project was one week late coming on-line and was under budget by more than one million dollars. Two years were required for completion. Of the $7.2 million allocated for the project, $2.3 million was spent on retrofitting.

---

* A mill equals $0.001.

Weyerhauser engineers actually operate the plant, which runs continuously, except over the fourth of July weekend and Christmas week when the entire plant is shut down for maintenance. Although the three boilers could supply one million pounds of steam per hour, enough to run the 52 megawatt steam turbine plant at full capacity, the turbines, running almost continuously, are actually used only at 50 to 60 percent capacity. The fact that the utility produces electricity at a cost of 18 mills per kwh is due to the low cost of wood waste fuel which Weyerhauser buys, the favorable economics of cogeneration, and the efficiency of the managers.

Weyerhauser is responsible for purchasing wood waste (not enough is generated on-site). EWEB pays Weyerhauser for the fuel at cost ($0.33/MMBTU for the wood waste; more, of course for oil) and at a rate of 4500 BTU per kwh. But as the heat rate is actually only 4200 BTU/kwh, this payment rate gives an additional incentive to Weyerhauser to produce steam for power generation. A further incentive is a 1.6 mill per kwh boiler operation and maintenance payment to Weyerhauser.

Similar utility industry partnerships could be founded all across the country with existing or planned boilers using wood, coal, oil, gas, or whatever. Instead of the initiative coming from a crisis like a drought, however, it should come from utilities seeking to cut their costs. PURPA encourages utilities to consider the potential for cogeneration before constructing any new capacity. State public utility commissions (PUCs) should require this examination. Since old ideas die hard, it is likely the utilities will resist the move from their monumental central stations to distributed cogenerators. Yet every indication is that power from cogeneration, either new or retrofit, will be more reliable and less expensive. PUCs would do well to hire independent consultants to assist in their evaluations. As for the federal government's role, the investment tax credit of 10 percent should be increased to 20 or 30 percent and extended beyond the 1983 expiration date. Alternatively, accelerated tax depreciation schedules could be afforded industries and utilities for investments in cogeneration depending on local financial needs. Technical assistance and consultation should be made available by the Department of Energy to PUCs and to interested investors. Third party cogeneration corporations should be allowed to flourish wherever industries and utilities fail to take advantage of an opportunity for cogeneration.

10.4.6. Research and Development

Robert H. Williams[5] makes a strong case for the need for cogeneration systems to produce a high proportion of electricity in relation to steam. Gas turbines do this best. A steam turbine system will generate only five megawatts of electricity from a 100,000 pound per hour steam load; a gas turbine system can satisfy that steam load but at the same time furnish 20 megawatts of power. The respective fuel savings from cogeneration are on the order of 25 and 50 percent compared with separate production of steam and electricity. Thus, the gas turbine, which may burn either oil or gas, is the most efficient system.

A system which uses solid fuels in a gas turbine system is the Fluidized Bed Combustor. Two approaches may be taken to fluidized bed combustion. Pressurized (PFBC) systems can probably be developed to use the hot combustion gases from coal or biomass to turn turbo-generators directly. Steam for industrial process use may be generated by constructing water pipes in the bed itself (the steam thus produced may also be used to generate electricity before process application) or by dumping the exhaust gases from the gas turbo-generators into a waste heat boiler. The atmospheric (AFBC) system, so called because the combustion chamber is at atmospheric pressure, generates steam for electric turbines and process applications through water tubes in the bed. Closed cycle gas turbines can be applied to AFBC systems which use helium in place of water, and these have electric to steam ratios midway between gas and steam turbines. A closed cycle AFBC system might generate 10 megawatts from a 50,000 pound per hour steam load.

Neither PFBCs nor closed cycle FBC systems are readily commercially available, unfortunately. AFBC systems for coal* may be on the market in a few short years, but the closed cycle gas systems† are further away. Pressurized FBC heat engines are perhaps ten years away. Technological optimists point out that the American Electric Power Company (AEP) and Stal-Laval, Ltd., of Sweden are planning a large PFBC plant in the near future, and that a prototype atmospheric system is already built in Rivesville, West Virginia. The former will certainly face a hostile life, however, for the erosive potential that coal combustion products carry for turbine blades is fierce. Operations above 1000°F may be impossible without metallurgical improvements. There is an urgent need to solve the problems of these systems while the demand for electric capacity is slack.[5,20]

10.4.7. New Institutions for Cogeneration

A key issue is preventing locking up of cogeneration capacity with the construction or retrofitting of industries' steam capability with less than optimal systems. Coupled with this issue is how to take advantage of the abundance of steam in relatively small industries. An Industrial Energy Park may be plausible concept to be applied toward these ends.

---

* Steam-only FBC systems for wood are readily available. Sand is the medium.
† Open cycle systems (air) may work as well as closed cycle helium.

Half the industrial expansion in the U.S. over the last few years has occurred in the sun belt. Much of this construction has occurred in industrial parks which offer to new industries not only utilities but also low-cost financing through the availability of low-interest tax-free municipal development bonds. Briefly, the Industrial Energy Park idea involves adding power and steam generated on-site as one of the services of such settings. Industries may find that assurance of an adequate energy supply is as vital as energy price. Moreover, on-site cogenerating power plants consuming wood waste, municipal solid waste (MSW), or coal and exporting electricity to a utility grid can be more economical, even at low load factors, than on-site steam generation and purchased electricity. So attractive is this idea, in fact, that utilities or a third party might want to offer to finance the electrical generating end of the system with the industry (or industries, municipal, or, again, a third party) buying the steam portion. Alternatively, the utility could own the entire plant and sell steam as a utility. Contractual arrangement could be designed to protect the industries from unfair rate hikes. Steam loads for the selected rapidly expanding industries are such that many small industries could benefit from cogeneration as long as one core industry, a chemical, paper, food, primary metals, brewery, or other plant with a steady steam load can be collocated. Where acreage for new plants exists around existing plants with cogeneration potential (or excess steam), such a park can be created. The Eugene experiment shows that the low load factors which might exist during the first years of the park's operation would not be economically devastating.

On the other hand, creation of such parks could be evolutionary. The process steam generating half of a system could be operational initially as in the case of an AFBC, for example, minus the closed cycle gas turbines. Once the load had developed sufficiently, the gas turbines could be added. As an incentive for such a strategy, tax credits, accelerated depreciation schemes, or eligibility for municipal development bonds, could be offered on advanced systems (PFBC) which could easily be converted to cogeneration. These inducements would be lower than for those using cogeneration from the start, but should be adequate to assure that the industry would follow through and install the complete system.*

Cogeneration can save more than one BTU for every BTU consumed, in relation to separate steam and electric generation. Making it happen, however, will be no easy task, for imagination will be required on the part of some very conservative institutions. Tax incentives and marginal energy

---

* Alternatively, industries cogenerating could be given variances on DOE's coal conversion requirements to allow them to use oil or gas. Cogeneration is an excellent way to use oil and gas.

pricing would help. While some have suggested giving variances for co-generation, the strict enforcement of the Clean Air Act should be applied. A policy of allowing variances could ultimately cripple FBC development. But something on the order of a national commitment will be required to capitalize on the opportunity cogeneration offers.

10.5. Summing up the Savings

Growth in industrial energy consumption will continue if the economy continues to grow. The rate of energy growth, as we have seen, need not be coupled in a one to one relationship with growth in output. Such has not been the case for thirty years, even with declining real energy prices, and with increasing industrial energy prices, we expect energy productivity growth to continue, if not accelerate. It is now clear that we can have both energy conservation and economic growth. *In fact, without the former, we will not be able to sustain the latter.*

A few industries consume very large quantities of energy. The chemicals, paper, and iron and steel industries are particularly large energy users. Fortunately, cogeneration, heat recovery, appliance efficiency and good housekeeping can effect large savings in existing plants. New processes will bring very large savings.

Promoting the large savings that are possible will require public policy initiatives. If industries assign discount rates of 50 percent (two year payback) to energy conservation investments, then far less than the optimum will be achieved. The government has fallen short, we believe, in providing incentives for industrial energy conservation investments. For cogeneration investments, in particular, inducements should be increased. Public utility rate structures should be changed to reflect the cost of service. Such policy might mean charging the marginal cost of electrical generation to new customers, a measure the Public Utilities Regulatory Policies Act of 1978 encouraged. This step alone would be a powerful incentive for cogeneration. Research and development on new industrial processes, on the efficiency of industrial appliances, and on technology for cogeneration should be vigorously pursued.

Without nonmarket incentives, industrial energy demand could burgeon to 40 quads by 2010. With a steady effort, an industrial energy demand level of less than 30 quads in the year 2010 can be adequate to insure economic growth, employment, and energy for environmental protection, and yet be sustainable.[1]

Epilogue

Through the Straits

E.1. Principles of Energy Conservation

The energy future which we envision is not some specific scenario, but rather a future still dimly seen that unfolds by events yet to occur and by application of certain principles. These principles include:

o   use of energy as a means, not an end;
o   application of technical ingenuity and institutional innovation;
o   making investment decisions with clear signals of total long-run, marginal costs, and cost trends;
o   correcting distorted or inadequate market signals with policy instruments;
o   internalizing to the extent possible the national security, human health, and environmental costs of using energy;
o   investment in energy supply and utilization research and development by both the public and private sectors;

o   providing consumer equity not with energy price controls, but with other means;

o   increasing cognizance of world conditions and needs, with special regard for international security and charity for the special needs of poor nations.

These principles, we hope, have been implicit in our exploration of the U.S. energy quandary.

Throughout this book, we have consistently used the term conservation as synonymous with increased energy productivity. Conservation is not curtailment, for curtailment is the reduction of the amenities that energy helps provide, and conservation is the substitution of capital equipment or ingenuity for energy in the provision of amenities. We have also based our analysis on the assumption that several decades might be available for an orderly transition to a renewable, relatively low (though still very much higher than elsewhere in the world) energy future. An oil embargo or similar event, of course, could drastically reduce our access to foreign oil and would thus alter this assumption. Some have claimed that such an event would destroy our democracy, and/or plunge our economy into a woeful depression. We do not believe that our political system is so fragile or that our dependence is so great. After all, the U.S. does produce 15 percent of all the oil in the world. And we have survived worse traumata.

This is not to say that dependence on foreign oil is healthy, sensible, or without danger. Surely the humiliation of being increasingly and un-necessarily dependent on hostile countries for a vital resource is reason enough to end or greatly reduce that dependence.

We believe that our system is resilient enough to withstand a major interruption of oil imports without calamity, without the need for tyranny or some energy war, although there would be hardships for many. But in order to minimize the shock of yet another oil crisis with attendant shortages and price escalations, we must prepare now.

E.2. Surviving the 1980s

E.2.1. Gasoline, The Highest Priority for Emergency Curtailment

The U.S. imports about 5 million barrels of oil per day and consumes 6.5 million barrels of gasoline per day. In a real sense, gasoline consumption *is* our energy crisis. One out of every nine barrels of oil consumed in the world is burned in U.S. gasoline engines.

Gasoline consumption per capita in Europe averages only one-fourth that in the U.S. Yet Europe and Japan cling even more tenuously to the

Table E.1. Oil Movements through the Strait of Hormuz, 1979

Country	Oil received via Strait of Hormuz (million barrels per day)	Percentage of total supply
France	1.9	83
Japan	4.0	71
United Kingdom	0.7	39
U.S.	2.4	13
West Germany	0.9	31

Source: Reference 1.

Middle East oil lifeline. If the two-mile wide Strait of Hormuz, the mouth of the Persian Gulf, were closed, France would lose 83 percent of its total oil supply, Japan 71 percent, Great Britain 39 percent, and West Germany 31 percent (see Table E.1). The U.S. would directly lose 13 percent of its total oil supply but would lose more indirectly due to international agreements to share oil shortages. The fact that Afghanistan is only 400 miles from the Strait of Hormuz will not be lost on the prudent.[1]

Because the U.S. is currently so extravagant in gasoline consumption, a great deal of flexibility would be possible in meeting another acute oil crisis. A thirty-five percent reduction in oil imports, for example, could be met by cutting gasoline consumption by two million barrels per day. Translated into effects on the U.S. automobile operator, such a loss would mean an average reduction in weekly automobile use from 220 miles per week to 155 miles. The issue, once it is accepted that such a reduction is generally tolerable, becomes one of how to share equitably and manage effectively the shortfall.

A number of options could be chosen to deal with a gasoline shortage: free market pricing, gasoline lines, rationing, and taxation. These tools, along with oil import fees and quotas and public transit and ride-sharing options can also be applied to reduce oil demand in advance of an emergency as a preventative measure. The case for imposing reductions in gasoline consumption becomes very strong in light of the national security risks of high oil imports, and when it is realized that the doubling of oil prices during the shortfall caused by the Iranian revolution was due to a shortage of only about two million barrels of oil per day on the world market. For this reason, and because the U.S. obtains about two million barrels of oil per day from the Persian Gulf, the argument is persuasive that we should reduce gasoline consumption by one to two million barrels of oil per day in the immediate future.

The most realistic policy tools for reducing gasoline consumption in advance or in the event of a gasoline emergency are rationing, taxation, or a combination of free market pricing and a windfall profits tax. At issue is how to select the most effective and equitable measure. It can be persuasively argued that free market allocation, coupled with windfall profits taxation, would do the best job of managing a shortfall. However, we believe that one possible variation would be to institute a sizeable imported petroleum excise tax and directly apply the tax revenue to filling the strategic petroleum reserve.

E.2.2. The Problems with Rationing

Gasoline rationing would be very expensive and quite probably would not work. Gasoline ration coupons would essentially be a new currency equal in physical volume to that of U.S. paper currency. Only with advanced, computerized accounting methods not yet available could rationing be made fair and manageable.

The greatest problem would be the sheer magnitude of the requisite system. Even if coupons were printed in five gallon denominations, twenty *billion* coupons would have to be distributed over a year. There are only seven billion U.S. currency notes in circulation. Because gasoline coupons would be legally negotiable, they could not be mailed directly to consumers because they might be stolen. Checks redeemable for coupons would be mailed instead. Banks estimate that in order to handle coupon transactions they would have to double their staffs. The situation would be vastly different from rationing in World War II (which did not work well), when there was a national commitment to the war effort, when there were almost 100 million fewer cars, and when the need for curtailment was unquestioned.

Equity, often cited as a positive attribute of rationing, would be served poorly by coupon allocation. The chief difficulty is to find a solid basis for allocating ration entitlements. If one chooses automobile registrations, one favors those who own more cars. If one chooses drivers licenses, then one excludes 30 million adults who could but do not hold them and opens the opportunity for fraudulent procurement of drivers licenses plus an incentive for unlicensed adults to obtain licenses. (One of the authors possesses two valid drivers licenses and would presumably be entitled to a double share of coupons.) Since there is no common identity number on licenses among states, fraud could be widespread.

Rationing would probably not be accepted by the public because it would mean long lines in banks, coupon checks lost in the mail, fraud, and inequity. Pressure would be great to abandon rationing as soon as gasoline supplies in storage were rebuilt, even if the crisis were still lingering.

E.2.3. The Case for a Gasoline Tax

We have already expressed the urgent need to curtail gasoline use. A gasoline tax of $0.50 to $1.00 per gallon and rebated on a per capita basis is the best means of effectively reducing gasoline consumption.

A tax on gasoline would drive down demand as consumers responded to the higher perceived price of gasoline at the pump. Estimates differ as to price elasticity of gasoline, but much evidence indicates that a $0.50 tax would reduce consumption by .7 to one million barrels of oil per day. A higher tax, one that equaled market clearing level, would be needed to sharply curtail demand in a crisis.

The tax should be rebated, for a $0.50 per gallon fee would raise $50 billion in revenues annually and without rebates would be devastating to consumers, especially the poor. Rebating a $1.00 per gallon tax on a per capita basis would return each adult about $700 per year, since the average American adult consumes about 700 gallons of gasoline per year. Low income consumers average 400 gallons of gasoline consumption per year, so income would be distributed to them from upper income groups, who consume about 1000 gallons per year.

The mechanisms for rebating the tax could be managed through income tax withholding, made rebatable to those who do not pay taxes, or through the social security system. Or the tax revenues could be used directly to purchase liquid fuels and add them to the strategic reserve.

A gasoline tax, therefore, can be effective and equitable. It would be manageable and inexpensive relative to rationing. Its application in advance of another oil embargo could be of inestimable value.

E.3. The Shibboleths of Energy

The Ascent of Man need not end with the exhaustion of oil. Neither should an oil embargo cause our economy to collapse upon itself like some burnt out star. By focusing future energy policy on energy conservation, we can buy the time needed to effect a transition to sustainable and renewable energy resources and a manageable annual demand, but we must not panic in the event of some (inevitable) supply disruption. Committing all available financial, technical, and leadership resources to some crash energy supply solution without taking full advantage of long-term conservation would be a terrible mistake. We would deplete those scarce resources, lose precious time, destroy the health and lives of many humans, ruin vast areas of the natural world, and still fail to meet the challenge. *In every area of energy, conservation remains the cheapest, most productive, most reliable, fastest, and safest alternative.*

To be sure, vigorous work must continue on the supply side if we hope to maintain even current levels of production. But our principal focus should remain on the most promising opportunities, those on the demand side of the equation. To the extent that we can quickly reduce demand, our existing supplies will last longer and will therefore supply our needs for a more measured and less costly transition to new sources.

A large portion of this book has been dedicated to demonstrating that expectations of enormous growth in the demand for energy cannot withstand rigorous examination. The authors of most of the projections we reviewed in Part I failed to take into account profound demographic and economic shifts in our economy. Assumptions were made in these auguries which are no longer defensible. These, the shibboleths of energy, include the belief that the use of energy must grow in order for the economy to grow; that energy demand, even in the long term is strongly price-inelastic; and still others that population, gross national product, and labor productivity would grow at rates exceeding those of the last two decades. These projections fixed in the minds of many the need for thousands of new power plants, coal mines, and synthetic fuels facilities. Against these conclusions we have contrasted the far smaller economic and external costs of energy conservation options.

In the buildings sector, we have shown that it is possible to reduce energy demand in existing buildings by as much as fifty percent, and at a cost far below that of producing electricity from either existing or new electric power plants, or the cost of energy from any new source.

In the transportation sector, we have illustrated the technical and economic feasibility of cutting the use of gasoline in cars in half, in a shorter period of time, at a lesser cost, and with far more certainty than an equivalent amount of synthetic fuels could be produced. Energy savings of almost as great a magnitude, also with favorable costs, are possible in the transportation of freight.

We described the potential for substituting capital and technical ingenuity for energy use in the industrial sector. Cogeneration, heat recovery, and basic housekeeping practices can save many quadrillion BTU per year. We believe that we have indicated how U.S. energy demand in the year 2010 need be no greater than 90 quads and could be as low as 60 quads. Such an energy future could be replete with the amenities we now enjoy. It would allow for expected population growth, as well as a considerable amount of growth in individual income.

We have suggested ways of utilizing alternate energy forms, energy carriers which will minimize adverse health and environmental effects. But, primarily, we have shown that the price of any form of energy is going to exceed by far the cost of a number of options for saving energy, and that diligent application of these options will reduce energy demand to a manageable level.

It is obviously contrary to our self-interest to neglect energy conservation, but effecting it will require the highest commitment from governments, industry, and private citizens. Energy conservation is the revolutionary event necessary not only to enable the ordinary events to continue, but to allow the human race to climb to whatever goals of equity, creativity, peace, and freedom it will. But it is only the beginning, for we have much to set right.

References

Prologue

1. Wills, G., *Inventing America,* Doubleday, New York, N.Y., 1978.
2. Sampson, A., *Seven Sisters,* Viking Press, New York, N.Y., 1976.
3. Gibbons, J. H., *et al., Alternative Energy Demand Futures to 2010,* U.S. National Academy of Sciences, Washington, D.C., 1979; also reported in *Science,* **200,** 142–152 (1978).
4. Hutcheson, F., *Inquiry into the Origins of Our Ideas of Beauty and Virtue,* 1725.
5. "Historical Statistics of the United States," U.S. Bureau of Mines, U.S. Department of the Interior, Washington, D.C., 1974.

Chapter One

1. Rodgers, W. H., Jr., *Environmental Law,* West Publishing Co., St. Paul, Minnesota, 1977.
2. Wills, G., *Inventing America,* Doubleday, New York, N.Y., 1978.

3. Report by the Federal Power Commission, *The 1970 National Power Survey,* Part I, U.S. Government Printing Office, Washington, D.C., 1971.
4. Meadows, D. H., Meadows, D. L., Randers, J., and Behrens, W. W., III, *Limits to Growth,* Potomac Associates, Washington, D.C., 1972.
5. Ray, D. L., *The Nation's Energy Future,* U.S. Government Printing Office, U.S. Atomic Energy Commission, Washington, D.C., 1973.
6. "Understanding the National Energy Dilemma," The Center for Strategic and International Studies, Georgetown University, Washington, D.C., August, 1973.
7. Report to the Energy Policy Project of the Ford Foundation, *A Time to Choose: America's Energy Future,* Ballinger Publishing Company, Cambridge, Massachusetts, 1974.
8. Dupree, W. G., Jr., and Corsentino, J. G., *United States Energy Through the Year 2000* (Revised), Bureau of Mines, U.S. Department of the Interior, Washington, D.C., 1975.
9. *The National Energy Outlook—1976,* Federal Energy Administration, Washington, D.C., 1977.
10. *1977 National Energy Outlook,* Federal Energy Administration, Washington, D.C., 1977.
11. Whittle, C. E., *et al., Economic and Environmental Impacts of a U.S. Nuclear Moratorium, 1985-2010,* Institute for Energy Analysis, Oak Ridge Tennessee, September, 1976; The MIT Press, Cambridge, Massachusetts, 1978.
12. Gibbons, J. H., *et al., Alternative Energy Demand Futures to 2010,* U.S. National Academy of Sciences, Washington, D.C., 1979. See also *Energy in Transition: 1985-2010,* W. H. Freeman, San Francisco, 1980.

Chapter Two

1. Whittle, C. E., *et al., Economic and Environmental Impacts of a U.S. Nuclear Moratorium, 1985-2010,* Institute for Energy Analysis, Oak Ridge, Tennessee, September, 1976; The MIT Press, Cambridge, Massachusetts, 1978.
2. "Twenty-Ninth Annual Electrical Industry Forecast," *Electrical World,* McGraw-Hill, Chicago, Illinois, 1978.
3. Allen, E. L., and Chandler, William U., "Regional Impacts of a U.S. Nuclear Moratorium," Institute for Energy Analysis, Oak Ridge, Tennessee (ORAU/iea 76-11) September, 1976.
4. Ward, Barbara, and DuBois, R., *Only One Earth: Care and Maintenance of a Small Planet,* W. W. Norton, New York, N.Y., 1972.
5. Savitz, Maxine, and Hirst, Eric, "Technological Options for Improving Energy Efficiency in Residential and Commercial Buildings," in *Energy Conservation and Public Policy,* edited by John C. Sawhill, Prentice-Hall, New York, N.Y., 1979.
6. Khazzoum, Daniel, "Proceedings of a Workshop on Energy Demand Modeling," Stanford Research Institute, Stanford, California, 1976.
7. Chandler, William U., "A Comparison of Energy Demand Projections," Institute for Energy Analysis, Oak Ridge, Tennessee, 1977.
8. Gibbons, John H., *et al., Alternative Energy Demand Futures to 2010,* U.S. National Academy of Sciences, Washington, D.C., 1979.

Chapter Three

1. Makhijani, A., *Energy Policy for the Rural Third World,* International Institute for Environment and Development, London, England, 1976.

)
2. Eckolm, E. P., *Losing Ground: Environmental Stress and World Food Prospects,* W. W. Norton and Company, New York, N.Y., 1976.
3. Report to the Energy Policy Project of the Ford Foundation, Makhijani, A., and Poole, A., *Energy and Agriculture in the Third World,* Ballinger Publishing Company, Cambridge, Massachusetts, 1975.
4. Gibbons, J. H., and Chandler, W., "Conservation of Energy in Its Production and Use," United Nations Environment Program, *Annual Report,* Nairobi, Kenya, 1978.
5. Warner, D., "Bangladesh: Is There Anything to Look Forward To?" *Atlantic,* **242** (64) November, 1978.
6. Pannill, H. B., "Energy: An Ethical Prospective," presented to "Energy in Perspective Symposium, 1978," Randolph-Macon College, Ashland, Virginia, 1978.
7. Pynchon, Thomas, *Gravity's Rainbow,* Viking Press, New York, N.Y., 1973.
8. Rose, D. J., and Lester, R. K., "Nuclear Power, Nuclear Weapons, and International Stability," *Scientific American,* Scientific American, Inc., New York, N.Y., 1978.
9. Darmstadter, J., Dunkerly, J., and Alterman, J., *How Industrial Societies Use Energy: A Comparative Analysis,* Resources for the Future, Johns Hopkins Press, Baltimore, Maryland, 1977.
10. Long, T. V., *et al., An International Comparison of Energy, Labor, and Capital Use in Manufacturing Industries,* University of Chicago, Committee on Public Policy, summarized in *International Comparisons of Energy Consumption,* J. Dunkerly, ed., Resources for the Future, Washington, D.C., 1978.
11. Lonnroth, M., *et al., Energy in Transition,* Secretariat for Future Studies, The Swedish Institute, Uddevalla, Sweden, 1977.
12. U.S. Central Intelligence Agency, *The International Energy Situation: Outlook to 1985,* Washington, D.C., 1977.
13. Martin, W. F., ed., *Workshop on Alternative Energy Strategies, Energy Supply to the Year 2000,* MIT Press, Cambridge, Massachusetts, 1977.
14. Schipper, Lee, Book Review of *Workshop on Alternative Energy Strategies* for Bulletin of the Atomic Scientists, 1978.
15. Organization for Economic Cooperation and Development, *World Energy Outlook,* Paris, France, 1977.
16. *World Energy Outlook,* Exxon Corporation, Public Affairs Department, New York, N.Y., 1978.
17. Leontieff, Wassily, *et al., The Future of the World Economy,* Oxford University Press, New York, N.Y., 1977.

Chapter Four

1. *National Geographic,* National Geographic Society, Washington, D.C., 1978, **154** (5), 632.
2. *Fishing Facts,* Northwoods Publishing Company, Menomonce Falls, Wisconsin, November, 1976.
    Hubbert, M. King, "Energy Resources" in *Resources and Man,* National Academy of Sciences, Washington, D.C., 1969.
3. As reported in "Energy Resources to Meet Any Need," *Aware Magazine,* August, 1977.
4. Marland, Gregg, *A Random Drilling Model for Placing Limits on Ultimately Recoverable Crude Oil in the Coterminous United States,* Institute for Energy Analysis, Oak Ridge, Tennessee, 1976, and in *Materials and Society,* Pergamon Press, London, 1978.
5. Cram, Ira H., Chairman, Coordinating Committee, National Petroleum Council and American Association of Petroleum Geologists, 1971.

6. Martin, W. F., ed., *Workshop on Alternative Energy Strategies, Energy Supply to the Year 2000.* MIT Press, Cambridge, Massachusetts, 1977.

7. Krapels, Edward N., *Oil Supply Security,* Rockefeller Foundation, New York, 1979.

8. *Factbook on the Proposed Natural Gas Bill,* prepared by Citizen/Labor Energy Coalition, Energy Action, Inc., Washington, D.C., 1978.

9. "Natural Gas: United States Has It If the Price is Right," *Science,* **191,** February 13, 1976.

10. *Monthly Energy Review,* U.S. Department of Energy, Energy Information Administration, Washington, D.C., 1978.

11. Moody, J. D., and Geiger, R. E., *Petroleum Resources: How Much, Where, and When?,* MIT Press, Cambridge, Massachusetts, 1977.

12. *Oil and Gas Journal,* December, 1975.

13. Dorfman, M., Deller, R., Bebout, D., Loucks, R., "Evaluation of the Geopressured Geothermal Resources of the Texas Gulf Coast," in U.S. Department of Energy's *Energy and Mineral Resource Recovery,* Institute for Energy Analysis, Oak Ridge, Tennessee, 1977.

    Jones, F. H., "Natural Gas Resources of the Geopressurized Zones in the Northern Gulf of Mexico Basin," in *Natural Gas from Unconventional Geologic Sources,* prepared by the Board of Mineral Resources, Commission on Natural Resources, National Academy of Sciences, U.S. Energy Research and Development Administration, publication FE-7271-1, Washington, D.C., 1976.

    Papadopulos, S. S., Wallace, R. H., Jr., Wesselman, J. B., and Taylor, R. E., "Assessment of Geopressured Geothermal Resources in the Northern Gulf of Mexico Basin," in *Assessment of Geothermal Resources in the United States—1975,* D. F. White and D. L. Williams, eds., U.S. Geological Survey Circular 726, Washington, D.C., 1976.

14. *Enhanced Oil Recovery Potential In the United States,* Congress of the United States, Office of Technology Assessment, Washington, D.C., January 6, 1978.

Chapter Five

1. "Statistical Data of the Uranium Industry," U.S. Department of Energy, Grand Junction, Colorado, GJO-100, 1978.

2. "Information from ERDA," **2,** (44), November 12, 1976.

3. "AEC Gaseous Diffusion Plant Operation," United States Atomic Energy Commission, ORO-684, Oak Ridge, Tennessee, 1972.

4. "Draft Report of the Supply/Delivery Panel," Committee on Nuclear and Alternate Energy Systems, National Academy of Sciences, Washington, D.C., 1979.

5. Weinberg, A. M., personal communication, July, 1977.

6. Feiveson, Harold A., Von Hippel, F., and Williams, R. H., "An Evolutionary Strategy for Nuclear Power," Princeton University Center for Environmental Studies, Princeton, New Jersey, Report PU/CES 67, September, 1978.

7. Perry, A. M., and Weinberg, A. M., "Thermal Breeder Reactors," *Annual Reviews of Nuclear Science,* 1972.

8. Rudasill, C. L., *et al., Coal and Nuclear Generating Costs,* Electric Power Research Institute, Palo Alto, California, Special Report EPRI PS-455-SR, April, 1977.

9. Williams, R. H., "Industrial Cogeneration," in *Annual Reviews of Energy* **III,** 1978.

10. Chandler, W., "Coal Supply Policy: Issues Related to Extraction," Institute for Energy Analysis, Oak Ridge, Tennessee, 1977.

11. *Monthly Energy Review,* United States Department of Energy, Washington, D.C., June, 1979.

12. Nephew, E. A., "The Challenge and Promise of Coal," *Technology Review,* **76,** (2), Massachusetts Institute of Technology, Cambridge, Massachusetts, 1973.
13. Bohm, R. A., *et al.,* "The Economic Costs of Back to Contour Stripmining Reclamation: The TVA Massengill Mountain Experience," University of Tennessee Energy Environment, and Resources Center, Knoxville, Tennessee, 1977.
14. Hammond, A., and Metz, J., *Solar Energy in America,* American Association for the Advancement of Science, Washington, D.C., 1979.
15. "Energy from Biological Processes," Office of Technology Assessment, U.S. Congress, Washington, D.C., 1980.
16. Hunt, Herbert, Vice President of the Eugene Water and Electric Board, Eugene, Oregon, personal communication, September, 1978.
17. Flaim, Silvio J., *et al.,* "Economic Feasibility and Market Readiness of Solar Technologies" Draft Final Report, Solar Energy Research Institute, Golden, Colorado, September, 1978.
18. Steele, Robert V., *et al.,* "Synthetic Liquid Fuel Development: Assessment of Critical Factors," U.S. Energy Research and Development Administration, Washington, D.C., 1977.

Chapter Six

1. *Implications of Environmental Regulations for Energy Production and Consumption,* National Academy of Sciences, Washington, D.C., 1977.
2. *Oil Spills,* Oil and Special Materials Control Division, U.S. Environmental Protection Agency, Washington, D.C., 1975.
3. *Environmental Quality,* The Seventh Annual Report of the Council on Environmental Quality, U.S. Government Printing Office, Washington, D.C., 1976.
4. Lave, L. and Seskin, E. P., *Air Pollution and Human Health,* Resources for the Future, Washington, D.C., 1977.
5. *Air Pollution Primer,* National Tuberculosis and Respiratory Disease Association, New York, N.Y., 1971.
6. Nephew, E. A., and Spore, R. L., *Costs of Coal Surface Mining and Reclamation in Appalachia,* ORNL-NSF-EP-86, Oak Ridge National Laboratory, Oak Ridge, Tennessee, 1976.
7. Rodgers, W. H., Jr., *Environmental Law,* West Publishing Company, St. Paul, Minnesota, 1977.
8. Allen, E. L., and Chandler, W., *Regional Impacts of a U.S. Nuclear Moratorium,* ORAU/IEA (M)-76-11, Institute for Energy Analysis, Oak Ridge, Associated Universities, Oak Ridge, Tennessee, 1976.
9. Chandler, W., Federow, H. L., Poole, A. D., Rotty, R. M., and Tompkins, P. C., *Economic and Environmental Implications of a U.S. Nuclear Moratorium, 1985–2010,* Volume II, Institute for Energy Analysis [ORAU/IEA (M)-76-13], Oak Ridge, Tennessee, 1976.
10. Ackerman, B. A., Ackerman, S. R., Sawyer, J. W., and Henderson, D. W., *The Uncertain Search for Environmental Quality,* The Free Press, New York, N.Y., 1974.
11. *Rehabilitation Potential of Western Coal Lands,* National Academy of Sciences, Ballinger Publishing Company, Cambridge, Massachusetts, 1974.
12. Bohm, R. A., Gibbons, J. H., Minear, R. A., Moore, J. R., Schlottman, A. M., and Zwick, B., *The Economic Impact of Back to Contour Reclamation of Surface Coal Mines in Appalachia: The TVA Massengale Mountain Project,* Appalachian Resources Project, Environment Center, University of Tennessee, Knoxville, Tennessee, 1976.

13. Brown, R., and Witter, A., eds., "Health and Environmental Effects of Coal Gasification and Liquefaction Technologies," DOE/HEW/EPA-03. Workshop Summary and Panel Reports for the Federal Interagency Committee on the Health and Environmental Effects of Energy Technologies, Departments of Energy and Health, Education, and Welfare, and the Environmental Protection Agency, prepared by the MITRE Corporation, Metrek Division, McLean, Virginia, 1979.
14. *Proceedings of the 1977 Oil Spill Response Workshop,* U.S. Fish and Wildlife Service, U.S. Department of the Interior, Washington, D.C., 1977.

Chapter Seven

1. Gregg Marland, Staff Geologist, Institute for Energy Analysis, personal communication, October, 1979.
2. Dupree, W. G., Jr., and Corsentino, J. G., *United States Energy Through the Year 2000* (Revised), Bureau of Mines, U.S. Department of Interior, Washington, D.C., 1975.
3. Whittle, Charles E., *et al., Economic and Environmental Impacts of a U.S. Nuclear Moratorium, 1985–2010,* The MIT Press, Cambridge, Massachusetts, Second Edition, 1979.
4. Gibbons, John H., *et al., Alternative Energy Demand Futures to the 2010,* U.S. National Academy of Sciences, Washington, D.C., 1979.
5. Benneman, John, "Bioconversion: An Assessment," Electric Power Research Institute, Palo Alto, California, 1978.
6. "Energy from Biological Processes," Office of Technology Assessment, U.S. Congress, Washington, D.C., 1980.
7. Kelly, Henry, *et al.,* "Application of Solar Technology to Today's Energy Needs," U.S. Congress, Office of Technology Assessment, Washington, D.C., 1979.
8. Flaim, Silvio J., *et al.,* "Economic Feasibility and Market Readiness of Solar Technologies," Draft Final Report, Solar Energy Research Institute, Golden, Colorado, September, 1978.
9. Steele, Robert V., *et al.,* "Synthetic Liquid Fuels Development: Assessment of Critical Factors," U.S. Energy Research and Development Administration, Washington, D.C., 1977.
10. Rotty, Ralph M., "Thermodynamics and Energy Policy," Institute for Energy Analysis, Oak Ridge, Tennessee, 1978.
11. *Implications of Environmental Regulations for Energy Production and Consumption,* U.S. National Academy of Sciences, Washington, D.C., 1977.

Chapter Eight

1. Gibbons, J. H., *et al., Alternative Energy Demand Futures to 2010,* U.S. National Academy of Sciences, Washington, D.C., 1979.
2. Hirst, E., and Kurish, J. B., *Residential Energy Use to the Year 2000: A Regional Analysis,* Oak Ridge National Laboratory, Oak Ridge, Tennessee, 1977.
3. Hutchins, P. F., Jr., and Hirst, E., *Engineering–Economic Analysis of Single-Family Dwelling Thermal Performance,* ORNL/CON-35, Oak Ridge National Laboratory, Oak Ridge, Tennessee, 1978.
4. *The National Energy Act,* Department of Energy, Office of Public Affairs, Washington, D.C., 1978.

5. Naismith, N. C., *et al., Residential Energy Conservation, Volume 1,* Congress of the United States, Office of Technology Assessment, Washington, D.C., July, 1979.
6. McCaughey, J., *The Energy Daily,* November 29, 1979.
7. Flaim, Silvio J., *et al.,* "Economic Feasibility and Market Readiness of Solar Technologies," Draft Final Report, Solar Energy Research Institute, Golden, Colorado, September, 1978.
8. Hirst, E., and Jackson, J., *Energy 2,* **131,** 1977.
9. Penoyar, W. E., and Williams, F. E., "Survey of U.S. Residential Insulation Industry Capacity and Projections for Retrofitting U.S. Housing Inventory," *Construction Review,* U.S. Department of Commerce, Washington, D.C., 1977.
10. Rosenfeld, A. H., Wall, L. W., Deh, T., Gadgill, A. J., and Lilly, A. B., "Conservation Options in Residential Energy Use: Studies Using the Computer Program Twozone," Lawrence Berkeley Laboratory Report LBL-5926, Berkeley, California, 1977.
11. "29th Annual Electrical Industry Forecast," *Electrical World,* New York, N.Y., September 15, 1978.
12. Dubin, F. S., and Long, C. G., Jr., *Energy Conservation Standards,* McGraw-Hill, New York, N.Y., 1978.
13. Hise, E. C., and Holman, A. S., "Heat Balance and Efficiency Measurements of Central, Forced Air, Residential Gas Furnaces," Oak Ridge National Laboratory (ORNL/NSF-EP-88), Oak Ridge Tennessee, 1975.
14. Jackson, J. R., Cohn, S., Cope, J., and Johnson, W. S., *The Commercial Demand for Energy: A Disaggregated Approach,* ORNL/CON-15, Oak Ridge National Laboratory, Oak Ridge, Tennessee, 1978.
15. Hoskins, R. A., and Hirst, E., *Energy and Cost Analysis of Residential Water Heaters,* Oak Ridge National Laboratory, Oak Ridge, Tennessee, 1977.
16. O'Neal, D., Carney, J., and Hirst, E., *Regional Analysis of Residential Water Heating Options: Energy Use and Economics,* Oak Ridge National Laboratory, Oak Ridge, Tennessee, 1978.
17. Holman, A. S., and Brantley, V. R., *ACES Demonstration: Construction, Start Up, and Performance Report,* Oak Ridge National Laboratory, Oak Ridge, Tennessee, 1976.
18. Bedinger, A. F. G., and Bailey, J. F., "Performance Results and Operating Experience of the UT-TVA Solar House (1976–1978)," Environment Center, University of Tennessee, Knoxville, Tennessee, 1978.
19. Kesey, Ken, *One Flew Over the Cuckoo's Nest,* Viking Press, New York, N.Y., 1962.
20. Williams, R. H., personal communication, Princeton University Center for Energy and Environmental Studies, 1978.
21. Hoskins, R., and Hirst, E., "Energy and Cost Analysis of Residential Refrigerators," Oak Ridge National Laboratory (ORNC/CON-6), Oak Ridge, Tennessee, 1977.
22. U.S. President's Council on Environmental Quality, "The Good News About Energy," Washington, D.C., 1978.
23. Sant, R. W., "Coming Markets for Energy Services," p. 6, *Harvard Business Review,* May–June, 1980.
24. Gibbons, J., "Long-Term Research Opportunities," Chapter 8 of *Energy Conservation and Public Policy,* edited by John C. Sawhill, Prentice-Hall, 1979.

Chapter Nine

1. Hemphill, R., in *Energy Conservation and Public Policy,* edited by John C. Sawhill, Prentice-Hall, Englewood Cliffs, New Jersey, 1979.
2. Grey, J., Sutton, G. W., and Zlotnick, M., "Fuel Conservation and Applied Research," *Science,* **200** (4338), April 14, 1978.

3. Motor Vehicles Manufacturers Association, *MVMA Motor Vehicle Facts and Figures—'79,* Washington, D.C., 1979.
4. Kummer, J. T., "The Automobile as an Energy Converter," *Technology Review,* February, 1975.
5. Gibbons, J. H., *et al., Alternative Energy Demand Futures to 2010,* U.S. National Academy of Sciences, Washington, D.C., 1979.
6. Difiglio, C., "Cost Effectiveness and Analysis of 1985 Mandatory Fuel Economy Standards," Technical Memorandum, U.S. Department of Energy, Policy, and Evaluation, Washington, D.C., January 22, 1979.
7. Stevenson, R. R., *et al., Should We Have A New Engine: An Automobile Systems Evaluation,* Jet Propulsion Laboratory, California Institute of Technology, Pasadena, California, August, 1975.
8. Brown, L. R., Flavin, C., and Norman, C., *Running On Empty: The Future of the Automobile in a Oil Short World,* W. W. Norton, New York, N.Y., September, 1979.
9. Heywood, J. B., "Alternative Engines and Fuels: A Status Review and Discussion of R & D Issues," Sloan Automotive Laboratory, Massachusetts Institute of Technology, Cambridge, Massachusetts, November, 1979.
10. Wilson, D. G., "Alternative Automobile Engines," *Scientific American,* **239,** (1), July, 1978.
11. Argonne National Laboratory, "Environmental Development Plan for the Division of Transportation Energy Conservation, Energy Research and Development Administration," September, 1977.
12. Bullard, Clark, former Director of The Office of Conservation and Advanced Systems Policy, U.S. Department of Energy, Washington, D.C., personal communication, 1980.
13. Jenny, L., U.S. Congress, Office of Technology Assessment, Transportation Program, Washington, D.C., September, 1979.
14. Lewis, A., Assistant to the Vice President for Purchasing, Louisville and Nashville Railroad, Louisville, Kentucky, December, 1978.

Chapter Ten

1. Gibbons, J. H., *et al., Alternative Energy Demand Futures to 2010,* U.S. National Academy of Sciences, Washington, D.C., 1979.
2. Alcoa, Inc., personal communication with energy management, Alcoa, Tennessee, 1978.
3. Boerker, S., "Thermodynamic Quality of Energy Use in Industry," Institute for Energy Analysis, Washington, D.C., 1979.
4. Flaim, Silvio J., *et al., The Technical Feasibility and Market Readiness of Nine Solar Energy Technologies,* Solar Energy Research Institute, Golden, Colorado, 1979.
5. Williams, Robert H., *Industrial Cogeneration,* Center for Environmental Studies Princeton University, Princeton, New Jersey, published in Energy III, 1978.
6. Hammel, B. B., and Brown, H. L., "Energy Consumption and Conservation in Unit Processes," General Energy Associates, Inc., Cherry Hill, New Jersey, September 1, 1977.
7. Gordian Associates, Inc., for The U.S. Energy Research and Development Administration, "The Steel Industry, Rational Use of Energy Program Pilot Study," New York, N.Y., 1977.
8. Gordian Associates, Inc., for The U.S. Energy Research and Development Administration, "The Cement Industry, Rational Use of Energy Pilot Study," New York, N.Y., 1976.

9. U.S. Department of Energy, Industrial Energy Efficiency Improvement Program, "Annual Report Support Document Volume II," Washington, D.C., 1978.
10. Reid, Robert O., Chiogioji, Melvin H., "Technical Options for Improving Energy Efficiency in Industry and Agriculture" in *Energy Conservation and Public Policy,* edited by John C. Sawhill, Prentice-Hall, Englewood Cliffs, New Jersey, 1979.
11. *Kirk-Othmer Encyclopedia of Chemical Technology,* Volume 8, Wiley/Interscience, New York, N.Y., 1965.
12. *Upgrading Existing Evaporators to Reduce Energy Consumption,* Technical Information Center, Energy Research and Development Administration, Washington, D.C, 1977.
13. Boeing Engineers and Constructors, personal communication, Seattle, Washington, 1978.
14. Hill, C. T., and Overby, C. M., "Improving Energy Production Through Recovery and Reuse of Wastes," in *Energy Conservation and Public Policy,* edited by John C. Sawhill, Prentice-Hall, Englewood Cliffs, New Jersey, 1979.
15. Lehrer, Tom, National Brotherhood Week, RCA Records, New York, N.Y., 1952.
16. "Cogeneration," *Power Engineering,* **82,** 3, March, 1978.
17. Progress Report No. 1, "Energy Sufficient Industrial Park Feasibility Study," Energy Opportunities Consortium, Inc., Knoxville, Tennessee, August 1, 1978.
18. Prepared for Federal Energy Administration, *The Potential for Cogeneration Development in Six Major Industries by 1985,* Resource Planning Associates, Cambridge, Massachusetts, 1977.
19. Prepared for Department of Energy, *The Potential for Cogeneration Development in Six Major Industries by 1985,* Resource Planning Associates, Cambridge, Massachusetts, 1977.
20. Graves, R., Oak Ridge National Laboratory, personal communication, Oak Ridge, Tennessee, January, 1979.
21. Rudasill, C. L., *Coal and Nuclear Generating Costs,* EPRI Special Report No. PS-455-SR, Electric Power Research Institute, Palo Alto, California, 1977.

Epilogue

1. Chandler, W. U., and Gwin, H. L., "Gasoline Consumption in an Era of Confrontation," in *The Dependence Dilemma,* edited by Daniel Yergin, Harvard Center for International Affairs, Harvard University, Cambridge, Massachusetts, June 1980.

Index